Introduction to Mineral Deposits Geology

(Including Exploration, Mining & Mineral Economics)

Moses. A. Olade, PhD, MBA

Prof. Moses A. Olade, Ph.D., MBA
2301 Harbor Chase Dr.
Pearland, Texas 77584
United States of America
mosesolade48@gmail.com

ISBN 978-0-578-34652-6
Prescott Publishers
https://www.prescottbooks.com
Library of Congress Control Number: 2022904942

2620 Golden Creek Ln, Houston TX 77574

Preface

The study of mineral deposits is an important component of the undergraduate geology curriculum in most universities and polytechnics. This book is an introductory text on mineral deposits geology that includes mineral exploration, mining and mineral economics, adapted to the needs of undergraduate students at 300 and 400 levels who are studying the subject of Mineral Deposits as a full 3-unit course consistent with the syllabus recommended for geology by the Universities accreditation agencies. *Introduction to Mineral Deposits Geology* represents an invaluable and systematic description of metallic and non-metallic mineral deposits integrated with mineral exploration, mining and basic mineral economics. New concepts and developments in economic geology are integrated into the text.

The book is organized in eleven chapters: Chapter 1: Introduction to Mineral Deposits; Chapter 2: Formation of Mineral Deposits; Chapter 3: Magmatic Mineral Deposits; Chapter 4: Hydrothermal Deposits; Chapter 5: Sedimentary Mineral Deposits; Chapter 6: Metamorphism-Related Mineral Deposits; Chapter 7: Surficial Mineral Deposits; Chapter 8: Industrial Minerals and Rocks; Chapter 9: Mineral Exploration; Chapter 10: Mining of Minerals, and Chapter 11: Mineral Economics. A Glossary of Common Terms in Economic and Mining Geology is included at the end of the book. The book is intended solely for undergraduate students in Geology and Mining at Universities and Polytechnics, particularly in Africa where standard textbooks are scarce or too costly to be affordable. Other geoscientists and mining professionals may find the book helpful.

Dr. Moses A. Olade
Pearland, Texas, USA
April, 2022

Table Of Contents

Chapter 1: Introduction

1.1 What are Mineral Deposits?

Mineral deposits are natural accumulations of valuable minerals in defined locations within the Earth's crust. Mineral deposits are usually composed of an aggregate of either a mineral or group of minerals in unusually high concentrations considered to be of sufficient economic potential for extraction at the present time or in the immediate future. The degree of concentration of the mineral content is known as the "*grade*" while the quantity is called "*tonnage*". Mineral deposits occur as ore bodies with three-dimensional configuration representing their shape, lateral extent, size and thickness (Fig. 1.1). They are the main sources of mineral commodities used as raw materials for industrial production, and their exploitation and utilization have become an important mainstay of the world economy. Demand for economic minerals continues to increase significantly, just as the global population grows exponentially.

The valuable minerals found in mineral deposits are of three types: *metallic ores* (e.g, gold, copper), *non-metallic minerals* and *rocks* (e.g. barite, limestone, gemstone) and *mineral fuels* (e.g. coal, tar sands, bitumen) (see Fig. 1.2). Metals and other mineral products are obtained from the economic minerals by mining and mineral processing, followed by marketing of the commodities to manufacturers and other end users. Mineral economics has become an interesting field of study applied to mineral deposits involving the application of economic principles to the global production and consumption of mineral commodities, and their influence on mineral resource development and international trade.

Fig. 1.1 Outcrop of a chromite mineral deposit overlying c host rock

Mineral deposits are an integral part of their host rocks, and generally occur as one or several mineral or ore bodies that are extensive or continuous masses of *ore* distinctly distinguishable by form and character from the enclosing host rocks. These ores usually contain an assemblage of valuable minerals called *ore minerals* in association with one or more waste minerals called *gangue*. For example, a copper deposit within a granitic host rock may contain valuable minerals such as chalcopyrite and bornite with quartz and calcite as gangue minerals.

The following terms are commonly used to describe mineral accumulations according to their supposed economic potential:

Mineral occurrence is a concentration of an economic mineral that is considered valuable or that is of scientific or technical interest (or a locality or wherever a mineral of potential economic interest is found, without reference to quantity or quality).

Mineral deposit is a mineral occurrence of sufficient size and grade that it might, under the most favorable of circumstances, be considered to have some economic potential.

Ore Deposit is a mineral deposit that has been explored and evaluated and found to be of sufficient size, grade to be mined at a profit.

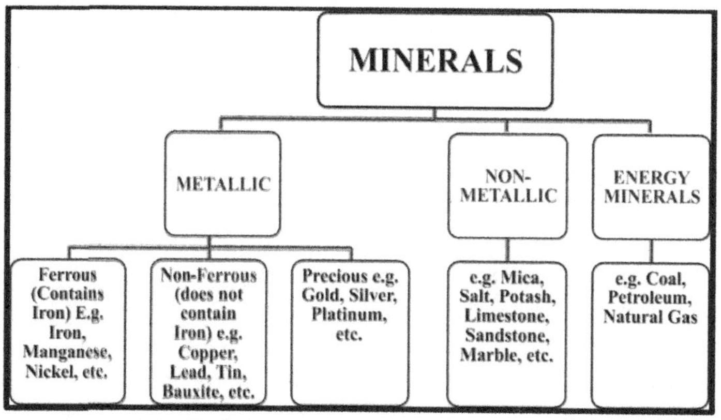

Fig. 1.2: Classification of economic minerals

Field observations of mineral discoveries usually begin with a “mineral showing” or "mineral occurrence" and progress with further study to "mineral deposit", and only rarely to "ore deposit," after detailed exploration and evaluation. The term *‘mineral deposit’* and *‘ore deposit’* are sometimes used interchangeably, but they are not exactly the same as shown in the preceding definitions. Although all ore deposits are mineral deposits, the reverse is not true because, not all mineral deposits are ore deposits. For a mineral deposit to become an ore deposit, it must satisfy not only the geological requirements such as sufficient tonnage and ore grade, but also satisfy other economic factors that ensure its viability for exploitation, such as metal prices, market conditions, transportation costs etc. Therefore, the difference between the two terms does have economic connotation. The term “mineral deposit” can be considered as a *geologic* term while “ore deposit” is an *economic* term.

1.2 Why Do We Study Mineral Deposits

Economic geology is the scientific study of mineral deposits. The importance of mineral deposits cannot be overemphasized because minerals, since ancient times, have been extracted from the bosom of the Earth and used for human benefit. History of human civilization and industrial development is closely linked to the use of economic minerals. Mineral deposits when exploited are the sources of the mineral commodities (metals, industrial minerals and mineral fuels) that are so indispensable to human existence and on which modern development is heavily dependent.

Minerals provide basic and essential raw materials for a lot of human needs including the production of a wide range of industrial and consumer goods, such as machinery, cars, computers, appliances; building and construction materials to develop, maintain and enhance our built environment and transport infrastructure, such as concrete roads, house, bridges; generation of electricity for use in the home, commerce and industry; transportation in the operation of vehicles, trains, aircraft and ships and agriculture to improve soil productivity and enhance food security. Several nations around the world also depend on minerals as a source of revenues for economic development through taxation and commodity export.

As the world's population grows exponentially and living standards increase globally, the demand for mineral raw materials continues to increase significantly. According to the United States Geological Survey (USGS), an average person in the United States consumed about 40,000 pounds of mineral resources in the year 2020 compared to 16,000 pounds in 2010, an increase of over 100% within a decade. However, mineral deposits are non-renewable resources, when they are mined, they cannot replace themselves except when new mineral discoveries are made. Still, those new discoveries will have to undergo exploration and evaluation for several years before they become mineable ore deposits. Consequently, to satisfy the world's hunger for mineral raw materials, there is a constant need to find new mineral deposits as the known reserves are depleted which requires the trained expertise of economic geologists and related professionals. The background analysis provided above leads us to the ultimate question of "*why do we study mineral deposits*"? We study mineral deposits so we can understand their basic geological characteristics and utilize such knowledge in identifying prospective areas of mineralization.

Economic geology is the scientific discipline focused on the study of mineral deposits so as to provide a good understanding of the distribution, geological characteristics, mode of occurrence and genesis of mineral deposits, and to apply such knowledge in the exploration and exploitation of mineral deposits. So far, economic geologists have been quite successful in finding more mineral deposits in known areas, deposits in new areas, and new types of deposits, and have continued to successfully expand and define new mineral reserves to satisfy human needs. However, the search for new mineral deposits is getting more difficult because the easily-found deposits that are exposed at the surface have mostly been discovered, and finding hidden deposits at depth will require more sophisticated and depth-penetrating technology, such as remote sensing. Mineral exploration efforts worldwide are confronted by many other challenges that include poor accessibility, political instability and inadequate security of life and property.

The study of mineral deposits and the search for new discoveries requires the use of several tools involving the knowledge of various aspects of geology and application of the principles and methods of various other fields of the geosciences, including geophysics, geochemistry and geostatistics. The basic understanding of the characteristics and genesis of mineral deposits requires the ability to identify ore minerals and mineralized host rock types, and undertake detailed geological mapping which requires a good knowledge of mineralogy, petrology, and stratigraphy. The understanding of structural features that control ore localization necessitates an adequate knowledge of structural geology and remote sensing. Mineral deposits are made up of minerals, both ore and gangue which both occur in ore bodies and host rocks. The ability to identify metallic and non-metallic minerals is crucial to ore discovery, ore recovery and mapping of alteration halos that are often associated with mineralization. The sub-disciplines of geosciences that are relevant to the study of economic geology include petrology, mineralogy, stratigraphy, structural geology, remote sensing, geochemistry, geophysics and geostastics.

Mineral deposits are usually defined or delineated by mineral exploration that uses geological and geochemical/geophysical methods, subsurface drilling and resource estimation through geostatistics and computer modeling to quantify economic ore bodies. A good knowledge of geochemistry, geophysics and geostatistics is required in the study of mineral deposits. An economic geologist who is involved in mineral exploration should also be familiar with field geological mapping and sampling methods, and the interpretation of geochemical and geophysical survey results. A good knowledge of mineral deposits is crucial to our mineral discovery efforts, and constitutes an important component in the training of geologists.

1.3 Classification of Mineral Deposits

Classification of mineral deposits is a process of grouping deposits into similar categories based on common or unifying features. Classification is a widely used procedure in geosciences to help simplify relationships and promote a better understanding of the substances being studied. Mineral deposits have been classified in many different ways using a number of parameters or concepts. Generally, mineral deposits are commonly classified into three main groups based on their mineralogical association: *metallic, non-metallic (industrial)* and *energy* mineral deposits. In addition, some geological criteria have been accepted and used for classification of mineral deposits including; (1) metal content (commodity); (2) shape or form of deposits (morphology); (3) host rock association; (4) ore forming processes, (5) conditions of deposition (temperature, pressure), (6) tectonic setting, and (7) origin (genesis). Each of these criteria may have its own merits or demerits, but classifying mineral deposits can be controversial because no two deposits are the same. Moreover, the formation of a deposit may be due to multiple processes superimposed on each other, thereby making it difficult to establish the exact ore-forming process. Table 1.1 summarizes the important classification criteria and identified groupings.

Table 1.1: Mineral Deposit Classification Criteria and Categories

	Classification Criteria	**Classification Categories**
1	Metal Content	Precious metals, Ferrous metals, Non-ferrous (Base) metals, Radioactive metals, High Technology metals
2	Shape and Form	Tabular, Tubular, Vein-type, Stratiform, Stratabound, Massive, Disseminated, Stockworks
3	Type of Host Rock	Mafic-Ultramafic, Alkaline-Peralkaline, Felsic plutonic/volcanic, Pegmatites, Sedimentary, Metamorphic, Placer-Paleoplacer
4	Ore Forming Processes	Disseminated, Stratiform-Layered, Massive sulfides, Pegmatitic, Hydrothermal, Chemical precipitates, Evaporites, Diagenetic, Biogenic, Alluvial, Metamorphic, Metamorphosed, Lateritic, Supergene
5	Conditions of Deposition (Temperature)	Telethermal, Epithermal, Mesothermal, Hypothermal
6	Tectonic Setting	Rift systems, Divergent plate margins, Convergent plate margins, Intracontinental settings. Orogenic belts, Aulacogens
7	Genesis (Geological Process)	Magmatic, Hydrothermal, Sedimentary, Metamorphic, Surficial

1.4 Nature of Mineral Deposits

Mineral deposits are unusual concentrations of economic minerals of sufficient size and grade within specific locations in the Earth's crust as a result of geological processes. These deposits exhibit certain unique characteristics some of which are evaluated as follows:

1.4.1 General Characteristics

1. Mineral deposits are known to occur both in rocks that form the continents (continental crust) and those that lie beneath the oceans (oceanic crust).
2. Mineral deposits are found in all geological environments including igneous, sedimentary and metamorphic.
3. Mineral deposits are an integral part of the host rocks in which they are found and are believed to have formed by the same or related geological process(es).
4. Mineral deposits have three-dimensional configuration that includes shape in plan and sectional view, continuity in strike and depth to represent volume, size and thickness.
5. Mineral deposits represent anomalous accumulations of particular elements/metals that are highly enriched at several times the normal concentration in ordinary rocks,
6. Mineral deposits are formed by both geochemically abundant and geochemically scarce elements.
7. Mineral deposits are composed of one or more ore minerals that are usually mixed with valueless minerals known as gangue.
8. Mineral deposits require some extraordinary energy source and/or unique geological processes to produce high metal enrichment.

1.4.2 Geochemical Abundance and Enrichment of Elements

Mineral deposits are products of geochemical enrichment of elements that accumulate in a defined location relative to the surrounding rocks. Nine major elements make up 99% of the Earth's crust These are the *geochemically abundant* elements: oxygen, silicon, aluminum, iron, calcium, sodium, magnesium, potassium and titanium that range in concentration from 0.5 - 10% in rocks. The rest 1% are the *geochemically scarce* elements such as Cu, Pb, Ni, and Au that occur in low concentrations in the parts per million (ppm) to parts per billion (ppb) range. These less abundant elements are those most commonly concentrated in mineral deposits, and have become valuable to modern society. These less abundant elements can be grouped into three: (a) *metals* (e.g. copper, zinc, gold)), (b) *semimetals* (e.g. antimony, bismuth), and (c) *non-metals* (e.g. phosphorus, nitrogen). They rarely form minerals of their own in common rocks, but occur in atomic substitution of other more abundant elements in the rock-forming minerals. For example, Cu, Zn or Ni may substitute for Fe or Mg in the structure of silicate minerals (e.g. olivine).

The high concentration of the geochemically scarce elements to form mineral deposits only takes place where a special geologic process has produced localized enrichment. Consequently, mineral deposits containing the scarce elements are less common or rare in nature, whereas ore minerals of the abundant elements, such as Fe can be found in many common rocks. Both geochemically abundant and geochemically scarce elements can be concentrated to form mineral deposits if the conditions are conducive.

The economic *concentration factor* of an element listed in Table 1.2 is the ratio of typical minimum economical ore concentration to average crustal concentration of the element, which in other words, is a measure of the level at which the element is enriched in a mineable mineral deposit relative to its abundance in the Earth's crust. Concentration factors can be calculated by dividing the average concentration of the metal or element in an ore deposit by its average crustal abundance. For example, the average crustal abundance of chromium (Cr) is about 0.01 wt %. Chromium ore can on the average be profitable if it contains 40 wt % chromium. The necessary concentration factor is therefore nearly 4,000 – chromium must be concentrated at least 4,000 times to be profitable. Table 2.1 lists the average crustal abundances and concentration factors for some of the important metals that are commonly extracted.

Concentration factors are highly variable and depend not only on the particular element but also on the type of deposit. For example, aluminum (Al) that has an average crustal abundance of 8.0 wt %, requires a concentration factor of 3 to 4 to form a mineable deposit, which means, containing 24 - 32 wt % Al to be considered economically viable. For some precious metals such as gold (Au) and silver (Ag), the level of concentration required to form a mineable deposit is usually in the thousands, and that may explain why gold and silver deposits are relatively rare. (Table 1.1). Thus, elements that occur in high abundance do not need a high concentration factor to make mining economical. whereas the scarce metals require greater concentrations to be profitably mined. The relatively abundant elements, such as iron and aluminum, are mined in many places around the world whereas the rarer elements, including tin, chromium, or lead, are mined n far fewer places.

Table 1.2: Concentration Factors of Metals in Mineral Deposits

Substance (Element)	Average Crustal Abundance	Typical Grades of Ore Deposits	Concentration Factor
Aluminum (Al)	8.0%	24 - 32%	3 to 4
Iron (Fe)	5.8%	40%	6 to7
Titanium (Ti)	0.86%	25 – 45%	25 to 100
Chromium (Cr)	0.0096%	40%	4,000 to 5,000
Zinc (Zn)	0.0082%	4%	300
Copper (Cu)	0.0058%	0.5 – 4%	100 to 200
Silver (Ag)	0.000008%	0.01%	~1000
Platinum (Pt)	0.0000005%	.0003	600
Gold (Au)	0.0000002%	0.0001 – 0.001	4,000 to 5,000
Uranium (U)	0.00016%	0.3	500 to 1000

Metals in Mineral Deposits

Mineral deposits are mostly exploited for the metals they contain. This is due to the importance of metals to human society for industrial production. About half of the chemical elements known in Nature possess some metallic properties. The term *metal*, however, is used for those chemical elements that possess two or more of the characteristic physical properties of metals (opacity, ductility, malleability, fusibility), and are also good conductors of heat and electricity. Approximately forty metals are produced from mining of mineral deposits. These important metals can be classified according to their geochemical association and uses as follows:

(1) *Precious Metals*: gold (Au), silver (Ag) and platinum group of metals (Pt, Ir, Rh, Pd).
(2) *Base (Industrial) Metals*: copper (Cu), lead (Pb), zinc (Zn), ¬ tin (Sn), aluminum (

(3) *High Technology Metals*: rare earths, (REE), lithium (Li), tantalum (Ta), zirconium (Zr)
(4) *Ferrous and Ferro-alloy Metals*: iron, manganese, nickel, chromium, molybdenum, tungsten, vanadium, cobalt.
(5) *Minor Metals and related Semimetals*: antimony, arsenic, beryllium, bismuth, cadmium, magnesium, mercury, selenium, tantalum, tellurium, titanium, zirconium, etc.
(6) *Fissionable Metals:* uranium (U), thorium (Th), radium¬ (Ra)

Metals are extremely important as mineral resources used in industry. For the industrial metals, *iron* is by far the most heavily used metal and fortunately is one of the most common. Iron is used to make iron and steel used in manufacturing and construction. *Aluminum* is the second most heavily used metal and is valued for its lightweight and strength (most airplanes are made of aluminum). *Copper, lead, zinc, nickel,* and cobalt are also important metals used widely in industrial applications such as electric wiring, alloys and electroplating. Chromium and tin are also very important metals used separately or as alloys in steel. The high-technology metals have become important in recent decades due to development of high-tech consumer electronics and green energy technology, such as electric vehicles and wind power.

Ore Grade of Mineral Deposits

Ore grade refers to the concentration of a desired metal or mineral in a mineral deposit. It is a measure of the level of enrichment compared to the surrounding rocks. Ore grade is a very important parameter in the evaluation of the exploitation potential of a deposit, The value of the metals or minerals in a deposit are weighed against the cost of extraction to determine whether it is of sufficiently high grade to be mineable. Ore grade is expressed as a percentage, e.g., 4% iron by weight (kilograms), but for precious metals that occur in very low concentrations, different units of measurement are used, such as ounces/ton or grams per kilogram or parts per billion (ppb).

Ore grade is an important parameter in the reporting of ore reserves particularly for the benefit of investors, and may exhibit considerable variation throughout a deposit. Moreover, there is a certain grade below which it is not profitable to mine a mineral even though the mineral is still present in substantial quantities. This is called the *cutoff grade* which can be better can be defined as the minimum grade required in order for a mineral or metal to be economically mined (or processed). Cut-off grades are usually calculated and used at various stages during exploration and mine development, starting with the ore reserve estimates, mine design and operation. Throughout the life of a mine, cut off grades may change due to external factors such as metal prices or new technology in mining.

1.4.3 Mineralogical Features

Mineral deposits are of economic value because of the useful minerals they contain. Such minerals are called *economic minerals* because they are of potential value. Economic minerals may be *metallic* or *non-metallic* (industrial). The metallic minerals are made up of one or more metals. Some important metallic minerals are magnetite (a source of iron), gibbsite (a source of aluminum), sphalerite (a source of zinc) and galena (a source of lead). The term '*ore*' is used to describe the assemblage of minerals found in mineral deposits that include *ore minerals* and *gangue minerals*. Table 1.3 is a list of the common metallic minerals and their formula.

Ore Minerals

Ore minerals are the useful minerals from which one or more metals can be extracted. Two factors determine whether a given mineral is suitable to be an ore mineral: (1) the ease with which a mineral can be separated from the gangue and concentrated which is based on the physical properties of the mineral (e.g. magnetism, density, flotation), and (2) the ease of releasing the metal from the other elements to which it is chemically bonded in the mineral. *Gangue* minerals are the valueless minerals found with the ore minerals that are usually discarded with the waste rock during mining and mineral processing. Both ore and gangue minerals are believed to be products of the same geological or mineralizing process. Some important ore minerals are: magnetite (a source of iron), bauxite (a source of aluminum), sphalerite (a source of zinc) and galena (a source of lead). A list of the most common ore minerals is presented in Table 1.3.

Ore minerals vary widely in their metal content. Ideal *ore minerals* contain 100% of the metal or commodity of interest, but are not very common. Native copper (100% Cu) is a good example, it occurs as native copper that is not common in economic concentrations. However, in reality, the most commonly mined ores may not be the ideal but are usually enriched in ore minerals that can be produced cheaply and the desired metals can be separated easily. In mode of occurrence, ore minerals usually form part of ores that contain one or more metals combined with anions or anion complexes, which are commonly oxygen, sulfur, sulfate and hydroxyl. Occasionally but rarely, these minerals may occur as a single or native metal as earlier discussed (e.g. native gold or native copper). Ore minerals can be generally classified into four groups: native elements, oxides, sulfides, and silicates (Table 1.4).

(1) ***Native elements*** are elements in native state that do not occur in combination with other elements. They include metals, semi-metals and non-metals. Only two metals, - *gold* and *platinum*, are found principally in their native state, although silver, nickel, copper, iron and several other metals also occur in the native state, but only a few occurrences are large and rich enough to be ore deposits (e.g. the native Cu deposits in the Keweenaw basalts of Michigan, USA). Native gold, silver, and copper may contain small amounts of other elements. Graphite and diamond made of carbon, and sulfur are examples of nonmetallic native elements.

(2) ***Sulfides*** are minerals composed of a combination of metal and sulfur. They constitute the largest group of ore minerals. The following metals copper, lead, zinc, nickel, molybdenum, silver, arsenic, antimony, bismuth, cobalt, and mercury; all form sulfide ore minerals in mineral deposits.

(3) ***Oxides and Hydroxides:*** These are minerals that combine metal and oxygen. The ore minerals of the most abundant metals such as aluminum, iron, manganese, and titanium belong to this group. Other less abundant metals, chromium, tin, tungsten, tantalum, niobium, and uranium also occur as oxides and hydroxides. Difference between oxides and hydroxides is the temperatures at which they form and are stable. Hydroxides exist in low-temperature environments and are commonly products of alteration and weathering.

(4) ***Silicates*** although common, are generally very rare ore minerals Exceptions are garnierite $(Ni,Mg)_6Si_4O_{10}(OH)_8$, a clay-like mineral that is a major ore mineral in Ni laterites; and grunerite ($Fe_7Si_8O_{22}(OH)$, an iron-silicate mineral that occurs in lateritic iron ores. These silicate ore minerals are of secondary origin produced by weathering.

Table 1.3: Some Common Ore Minerals and Their Formula

	Metal	Minerals	Formula
1	Al	gibbsite boehmite diaspore	$Al(OH)_3$ $AlO(OH)$ $AlO(OH)$
2	Fe	magnetite hematite goethite siderite pyrite	Fe_3O_4 Fe_2O_3 $FeO(OH)$ $FeCO_3$ FeS_2
3	Cu	copper chalcopyrite bornite chalcocite covellite	Cu $CuFeS_2$ Cu_5FeS_4 Cu_2S CuS
4	Ni	pentlandite garnierite	$(Ni,Fe)_9S_8$ $(Ni,Mg)_3Si_2O_5(OH)_4$
5	Zn	sphalerite wurtzite zincite franklinite	ZnS ZnS ZnO $ZnFe_2O_4$
6	Mn	hausmannite polianite pyrolusite cassiterite	Mn_3O_4 MnO_2 MnO_2 SnO_2
7	Cr	chromite	$FeCr_2O_4$
8	Ag	gold calaverite	Au $AuTe_2$

Table 1.4: Common Ore Mineral Groups

Group	Ore Minerals	Element
Native Metals	Gold Platinum Copper Silver	Au Pt Cu Ag
Native Semimetals	Arsenic Antimony Bismuth	As Sb Bi
Native Non-metals	Carbon Sulfur	C -Graphite, Diamond S
Sulfides	Pyrite Chalcopyrite Molybdenite Sphalerite Galena Cinnabar Stibnite	FeS2 CuFeS2 MoS2 ZnS PbS HgS SbS2
Oxides	Corundum Magnetite	Al2)3 Fe3O4

	Hematite Pyrolusite Rutile Cassiterite Chromite	$Fe2O3$ $MnO2$ $TiO2$ $SnO2$ $FeCr2O4$
Hydroxides	Gibbsite Goethite Manganite	$Al(OH)3$ $FeO(OH)$ $MnO(OH)$
Silicates	Garnierite Grunerite	$(Ni,Mg)_6Si_4O_{10}(OH)_8$, $(Fe_7Si_8O_{22}(OH)$,

Gangue Minerals

Gangue minerals are the non-valuable minerals associated with ore minerals in ore bodies. The most common gangue minerals are silicates and carbonates, including quartz, feldspar, mica, chlorite, clays, calcite or dolomite, tourmaline and garnets. They are found closely associated with certain ore minerals. For example, quartz is the gangue in most vein gold deposits. Minerals that are produced by hydrothermal alteration in epigenetic deposits may constitute part of a gangue assemblage. Examples of such minerals include sericite, biotite, kaolinite, chlorite and epidote in porphyry copper deposits. Contact metasomatic deposits formed in carbonate rocks may be associated with a unique assemblage of calc-silicate minerals called "skarn", including wollastonite, diopside, forsterite, grossularite and epidote.

Non-Metallic Minerals

A group of economic minerals found in mineral deposits are not metals or fuels and known as *non-metallic minerals*. They are valued for their physical and chemical properties. They are used most commonly for industrial or construction purposes, and also serve as sources of valuable non-metallic elements such as potassium, sodium, phosphorus and sulfur. Non-metallic minerals (not including rocks which are aggregates of minerals) can be classified into two groups: *industrial minerals* (e.g. barite, feldspar), and *gemstones* (diamond, emerald, ruby). Table 1.5 is a list of the common non-metallic minerals (and rocks) and their groupings.

In recent years, the interest in industrial minerals (and rocks) has increased considerably due to the increasing contribution of this class of mineral deposits to the economy of many developed and developing countries. The rise in the exploitation of industrial minerals and rocks as a result of their widespread occurrence and uses has galvanized the efforts of geological exploration for these deposits. In developed countries such as the United States, the annual dollar value of industrial mineral production has since surpassed that for metals, and continues to grow rapidly. Industrial rocks and minerals are of great economic value as they serve as major raw materials in building construction and the manufacture of glass, abrasive, paper, chemicals, ceramics, metallurgical and agricultural products.

Table 1.5: Types of Non-Metallic Minerals

Industrial Minerals		**Gemstones**
Andalusite	Kyanite	Amethyst
Asbestos	Mica	Aquamarine
Barite	Olivine	Diamond
Bentonite	Magnesite	Emerald
Borates	Potash	Garnet
Clays	Quartz /Quartz Sand	Ruby
Corundum	Sillimanite	Sapphire
Feldspar	Talc	Topaz
Fluorspar	Trona (Soda Ash)	Tourmaline
Graphite	Wollastonite	Zircon
Gypsum	Zeolites	
Halite	Zircon/Zircon Sand	

1.4.4 Physical Attributes

Mineral deposits show certain physical attributes that distinguish them from other geological products. These include their mineral textures, mode of occurrence, size, shape and thickness.

Morphology

Most deposits show three-dimensional configuration which is reflected in their shape, size and lateral continuity within a variety of host rocks. The shape can range from regular such as in iron ore or coal deposits to extremely irregular as in disseminated gold and platinum deposits. Mineral deposits usually show either a *concordant* (parallel to) or *discordant* (cut across) relationship with the surrounding rocks which may indicate theis origin as either *syngenetic* (formed at same time as host rocks) or *epigenetic* (formed after the host rocks). Three types of regularly-shaped discordant ore bodies can be distinguished by shape: *isometric, tabular,* and *tubular. Tabu*lar *or* flat ore bodies such as sheets, veins and lenses, have two long dimensions and one short dimension, and usually appear at angles ranging from near horizontal to vertical.

Concordant ore bodies show conformity with the host rocks. They are usually *syngenetic* and show considerable development in two dimensions (parallel to bedding) and are limited in the third dimension. There are two types: stratabound and stratiform. *Stratabound* refers to any type of ore body that is restricted or confined to a particular part of the stratigraphic column or lies within a defined stratigraphic unit. They may be remobilized in forms that are discordant to the bedding. *Stratiform* **i**s used to describe a mineral deposit that is stratified or layered or occurs as a specific stratigraphic (or sedimentary) unit. Concordant ore bodies are found in a variety of sedimentary and igneous rocks, Sedimentary rock deposits such as limestone or phosphate beds, and chromite layers in mafic-ultramafic rocks are examples of stratiform deposits.

M*ode of occurrence* of a mineral deposit refers to its attitude or physical disposition. It is a descriptive term that connotes the general appearance and relationship of the mineral deposit to host rocks incorporating several characteristics that can be observed in the field. In general, magmatic deposits are usually massive, layered, lenticular, or pod-like in form, while sedimentary ore deposits are roughly tabular or stratiform. Hydrothermal deposits may occur as lodes, reefs, veins, veinlets, stringers, stockworks, and replacement bodies, while most other deposits may appear in a variety of modes including a blanket form for lateritic bauxite deposits.

The pattern of distribution of ore minerals in mineral deposits is one of the important physical attributes. Some ore minerals are scattered randomly (disseminated) or in stockworks of closely-spaced, multidirectional veinlets cutting the host rock and forming an interlacing network. Examples are copper, molybdenum and tin deposits in granitic and granodiorite rocks. Some mineral deposits occur as veins and veinlets occupying structural openings in rocks, such as fractures and faults.

Mineral deposits vary in *size* ranging from small veins of gold a few meters wide to large or "super-large" deposits extending for several kilometers such as iron ores or porphyry copper deposits. The *size* of a mineral deposit is an important factor in its economic viability. It is a geometric reflection of the quantity and spatial disposition of the ore body. Size distribution of mineral deposits has been studied extensively by economic geologists, and two distribution patterns have emerged: and are referred to as *fractal* and *exceptional.* A fractal distribution is associated with those deposits in which their sizes are dependent on the rate of accumulation within a defined space, such as a magma chamber or a fault zone. *Exceptional* size is observed in deposits that were formed by extreme or unique events in the Earth, such as meteoritic impacts that create massive space for ore accumulation, or mantle plume events that create large igneous bodies such as the Bushveld Igneous Complex of South Africa , or major climatic factors that produced exceptional conditions for the widespread deposition of mineral deposits such as banded iron formations during the early Precambrian or accumulation of thick lateritic bauxites along well preserved erosion surfaces, e.g. bauxite deposits in West Africa.

Mineral Textures

The textures of ores refer to the appearance, forms, and interrelationships of the minerals. The identification and interpretation of mineral textures can be very helpful in understanding the processes involved in the formation of mineral deposits. The nature of ore textures is usually determined by many factors, including the composition of the mineralizing fluids, the physical conditions of ore formation, and the nature of available space within the host rocks. There are quite a variety of textures observed in ore minerals, but they can be simply classified into three types: (1) *magmatic precipitation textures*, (2) *open- space filling textures* that permits the precipitation of minerals, and (3) *replacement textures* that are derived from changing the pre-existing conditions in the host rocks.

Magmatic precipitation textures are uniquely characteristic of magmatic ore deposits in which oxides and sulfides crystallize contemporaneously in the magma. The growth of ore minerals in silicate melts during the early phase of crystallization generally results in the development of euhedral to subhedral crystals because there is a little obstruction to the growth of mineral faces. *Cumulus textures* are developed when early formed crystals of ore minerals accumulate or settle from liquid melt to the bottom of a crystallizing magma. This is a texture reflecting accumulations of the same mineral in solid form during magmatic crystallization (Fig. 1.4a). Open space filling texture develops when hydrothermal solutions flow into open fractures. The deposition of layers of different minerals in the walls of a cavity leads to the development of what is known as *crustification*. Successive deposition of layers of crystals by hydrothermal solutions which are developed within the cavities, give rise to what is called *comb* structure or *colloform banding* which is common in low temperature deposits (Fig. 1.4b).

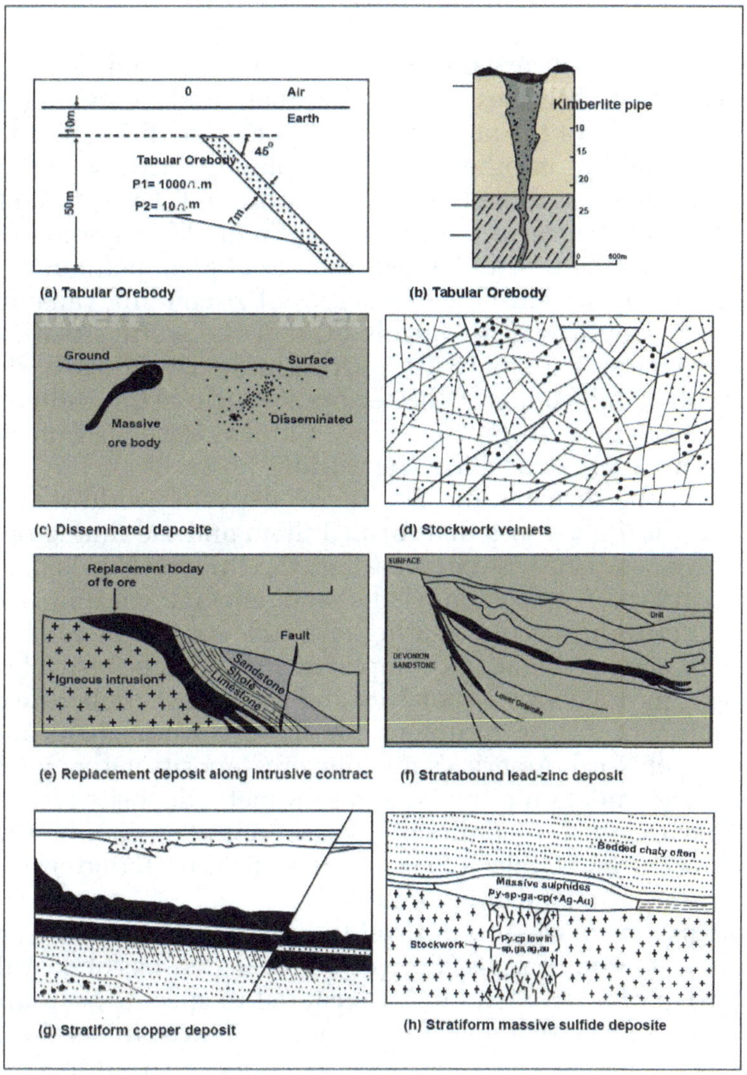

(a) Tabular Orebody (b) Tabular Orebody

(c) Disseminated deposite (d) Stockwork veinlets

(e) Replacement deposit along intrusive contract (f) Stratabound lead-zinc deposit

(g) Stratiform copper deposit (h) Stratiform massive sulfide deposite

Fig. 1.3: Morphology of mineral deposits

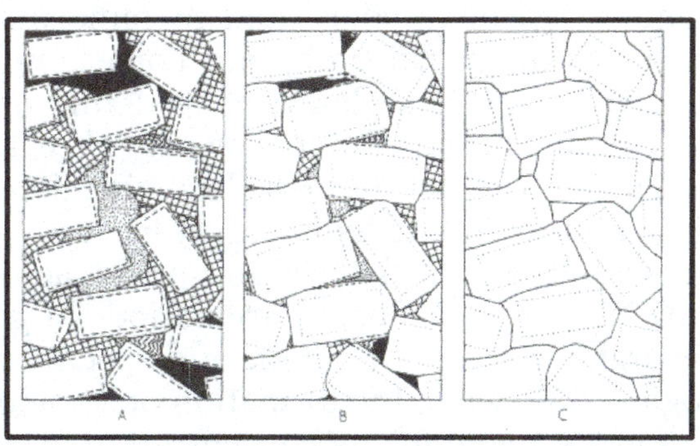

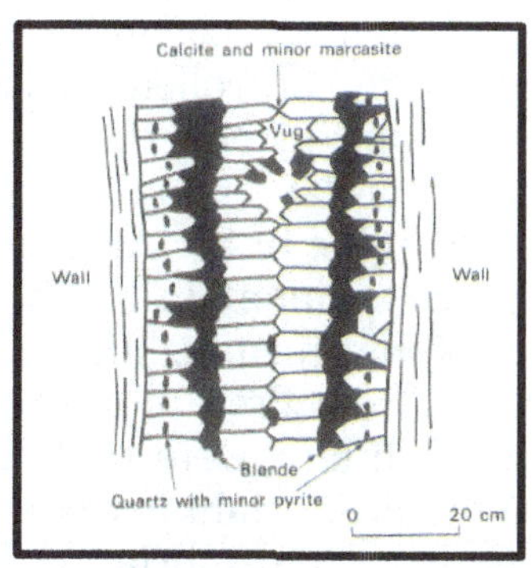

Fig. 1.4: (a)Cumulate texture magmatic ores (b) Open space filling and banding

1.5 Worldwide Distribution of Mineral Deposits

The distribution of mineral deposits around the world is uneven and seems to vary both in space and time due to the geological processes that formed them and the timing of such events. Some parts of the world contain most of the supply of certain commodities, and the minerals industry flourishes in some places and not in others. For example, in Africa where mineral resources are abundant, they are not evenly distributed across the continent. Over 70% of the known mineral reserves are confined to Southern Africa, out of which South Africa, arguably the leading mining country in the world, contains the world's largest reserves of gold, platinum group metals, chrome and manganese, as well as the second-largest reserves of zirconium, vanadium and titanium. By this generous endowment, South Africa controls 90% of the world's platinum, half the world's gold, and 75% of the chromium. Parts of Central Africa in the DR Congo and Zambia are also rich in mineral deposits of cobalt and copper. Other examples include the United States that produces 75% of the world's molybdenum from one mine, Climax Mine in Colorado, and China that accounts for 95% of the rare earth deposits in the world.

The differences in the distribution of mineral deposits worldwide can be attributed to different geology or geological processes that formed them and the timing of the occurrences. For example, in Southern Africa which was cited earlier, the mineral deposits are associated with a preponderance of Precambrian greenstone belts and ancient volcanic rocks that have been deformed during several orogenic events creating favorable environments for mineral deposition. Spectacular masses of mineralized igneous complexes such as the Bushveld in South Africa, the Great Dyke in Zimbabwe, the Palabora carbonatite and numerous kimberlite dykes have enhanced the mineralization potential of this subcontinent. Similar greenstone belts and associated rocks in Canada account for many of North America's metallic ore deposits. Other geological environments most conducive for the accumulation of metal and semimetal deposits are found near margins of continents or the former margins of continents where mountain building in orogenic belts and igneous activity have occurred. Other types of deposits are found in ancient rift systems, aulacogens and other intracontinental environments.

Economic geologists have tried to unravel the reasons for the patterns of variation in the spatial and temporal distribution of mineral deposits. Research has shown that mineral deposits of a certain metal or group of metals tend to be clustered in specific geographic areas or regions, known as *metallogenic provinces* or *mineral belts*. Some of these mineralized regions or districts are associated with specific metals that have been exploited for centuries, for example the base metal districts in the Kuperschiefer belt of north-central Europe. The spatial groupings of mineral deposits occur because geological and tectonic processes that involve mineral deposition, such as the formation of sedimentary basins, emplacement of igneous bodies, and metamorphism in orogenic belts tend to be controlled by larger processes of crustal evolution. For example, the Cretaceous Benue Trough in Nigeria is characterized by the abundance of lead-zinc deposits within a linear sedimentary basin formed by rifting during the Cretaceous breakup of Gondwanaland and opening of the Atlantic Ocean. The distribution of those lead-zinc deposits is linked to a stage in the continental rift evolution of that intracratonic sedimentary basin.

The *timing* of ore-forming processes also varies, and is known to spread across the geological time frame, from Precambrian to Recent. *Metallogenic epochs* are periods of geologic time during which conditions were particularly favorable for the formation of a specific type of mineral deposit. There are certain periods in geological time when the formation of some deposit types was prevalent. The Precambrian period was most notable for its concentration of mineral deposits. It is estimated that about 75% of known metallic mineral deposits are associated with areas underlain by Precambrian rocks. For example, a large proportion of gold and iron deposits are found within the Precambrian rocks

1.6 Life Cycle of Mineral Deposits

Mineral deposits usually pass through a life cycle starting with the discovery of the deposit through prospecting and/or exploration, except if it is an already known mineral occurrence or prospect. This is followed by preliminary and detailed exploration to define the extent and value of the ore body which leads to ore reserves estimation commonly using geostatistics to determine the size and grade of the deposit. The reserve estimates obtained and other economic and marketing factors are used to conduct a pre-feasibility study to determine the potential economic viability of the ore deposit. The next step after this is to conduct a full feasibility study to evaluate the operational costs, economic and financial viability, the technical and financial risks, and the profitability of the project.

At this point, the exploration or mining company makes the decision whether to move ahead and develop the mineral deposit into a mine or to "walk away" from the project. Such detailed evaluation is usually followed by mine design and planning for extracting the resource and to evaluate the economically recoverable portion of the deposit, the beneficiation and ore recoverability, marketability and payability of the ore concentrates, engineering concerns, milling and infrastructure costs, finance and equity requirements, and an analysis of the proposed mine from the initial excavation all the way through to reclamation. The proportion of a deposit that is economically recoverable is dependent on the enrichment factor of the ore body.

Once the feasibility analysis determines a given mineral deposit is worth being exploited, mine design and development of infrastructures begin so as to create full access to the ore body. The mine buildings and processing plants are built, and any necessary equipment is obtained. The operation of the mine to recover the valuable ore begins and continues as long as the company operating the mine finds it economical to do so. Once all the ore that the mine can produce profitably is recovered, the mine is closed and reclamation begins to make the land used by the mine suitable for future alternative use and environmentally safe. The life cycle of a mine from exploration to closure (cradle to grave) usually averages 30 years.

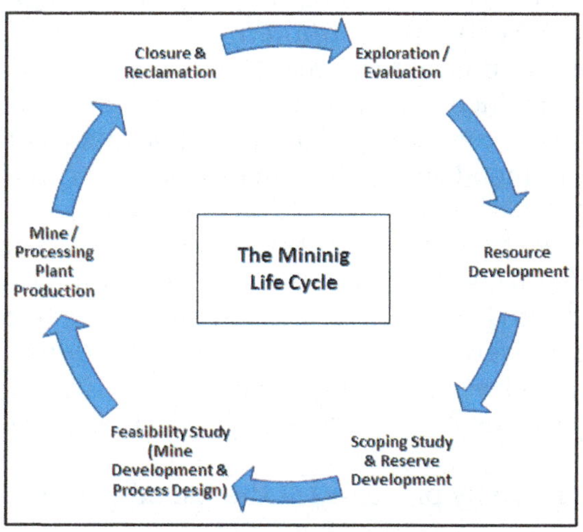

Fig. 1.4: Life cycle of ore deposits

The steps in the mineral deposit life cycle are outlined below and shown in Figure 1.4.

1. Prospecting and (discovery)
2. Exploration and Evaluation
3. Feasibility study
4. Mine planning
5. Mine development
6. Mine production
7. Processing/beneficiation/milling
8. Marketing
9. Closure and land reclamation

1.7 Definition of Common Terms in Mineral Deposits Geology

Several terms are commonly used in Economic Geology that require initial definition to assist the reader in the understanding of the subject matter.

Mineral occurrence--a concentration of a mineral that is considered valuable or that is of scientific or technical interest.

Mineral deposit– a localized concentration of economic minerals at places near the surface of the Earth where they can be extracted and converted to useful commodities.

Mineral showing – a mineral occurrence that is so small and may only be of geological interest but not necessarily of economic interest.

Mineral prospect - any promising mineral occurrence that has been partially developed by underground or by above-ground techniques, or by subsurface drilling.

Mineral processing - the process of extraction and concentration of economic minerals contained in ore.

Ore deposit - a mineral deposit that has been tested and found to be of sufficient size, grade, and accessibility to be mined at a profit.

Ore body - an extensive or continuous mass of ore and gangue distinctly distinguishable by form and character from the enclosing host rocks.

Ore - the valuable material extracted from an ore deposit. It is an aggregate of minerals from which one or more metals can be extracted *profitably.*

Orel reserves – the portion of mineral resources that is economically feasible to extract.

Ore mineral - a mineral from which one or more metallic elements can be extracted.

Ore grade - a measure of the concentration of a mineral or metal contained in an ore body

Cut-off grade - the minimum metal concentration or lowest permissible grade of ore that is considered minable at a profit,

Syngenetic –ores and minerals that formed at the same time as their host.

Diagenetic – ores emplaced during the compaction and consolidation of sediments

Epigenetic – ores emplaced into pre-existing rocks of any origin (e.g. veins, metasomatic

Hypogene – primary ores that were formed by ascending solutions.

Supergene – ore formation by descending meteoric water during weathering.

Endogenic – concentration caused by processes in the Earth's interior (magmatism or metamorphism)

Exogenic – concentration caused by processes at the Earth's surface (e.g sedimentation, weathering).

Disseminated ore - the scattering of ore minerals uniformly throughout host rock.

Massive ore – mineralization composed of uniformly a large proportion (>60%) of ore minerals
Host rock - the rock or rocks containing or surrounding a mineral deposit.
Tenor – the percentage of the ore mineral that is actual metal to be extracted. Often used as synonymous with ore grade.
Tonnage - the quantity of metal or useful mineral in a deposit. Usually expressed as the number of tons available at a specific average grade
Metallogeny - the study of the origin (genesis) of ore deposits, particularly in relation to tectonic activity.
Metallogenic province - an area of the earth's crust characterized by a particular assemblage of mineral deposit types.
Metallogenic epoch - a period of geological time favorable for the deposition of mineral deposits characterized by specific deposit types.
Stratiform – Mineralization that is conformable with host rocks and laid down like strata.
Stratabound – Mineralization restricted to a particular stratigraphic unit or interval but may be discordant in places.

Review Questions

1. What are mineral deposits? Briefly discuss some of their characteristics.
2. Describe your understanding of element concentration factors and how they relate to the concentration of the geochemically scarce elements.
3. What are ore minerals? Describe main types of ore minerals and highlight two factors that determine whether a mineral can be designated an ore mineral.
4. Define and differentiate the following pairs of terms:
    a. Mineral occurrence and Mineral prospect
    b. Mineral deposit and Ore deposit
    c. Metallic and Non-metallic minerals
5. Discuss the life cycle of mineral deposits. Explain when a mineral deposit becomes an ore deposit.

ORE MINERALS

NATIVE ELEMENTS

Gold

Silver

Copper with Malachite (green)

Platinum

Native Antimony

Native Sulfur

SULFIDES

Pyrite (FeS_2)

Galena (PbS)

Sphalerite (ZnS)

Magnetite (Fe_3O_4)

Cassiterite (SnO_2)

Ilmenite ($FeTiO_3$)

Hematite (Fe_2O_3)

HYDROXIDES

Goethite (Fe(OH)3

Gibbsite Al(OH)3

Manganite (MnO(OH)

SILICATES

Garnierite (Green Nickel)

Grunerite (Fe silicate)

NON-METALLIC MINERALS & ROCKS

Barite (Industrial Mineral)

Phosphate Rock

Sand (Construction Material)

Aquamarine

Diamond

Emerald

Chapter 2: Formation of Mineral Deposits

2.1 Introduction

Mineral deposits are formed because ore-bearing fluids of various types are able to *concentrate* and *transport* metals as ore minerals and *precipitate* them within favorable geological environments due to natural processes including decrease in temperature, and physical or chemical reactions that lead to precipitation or segregation of useful minerals. Sources of elements in mineral deposits can vary, including partial melting in the upper mantle and crust, and leaching or dissolution of elements from surrounding rocks. Metals can also originate from seawater and volcanic activity. Examples of ore-bearing fluids are magma, seawater, connate water and meteoric water, while examples of precipitating processes are the crystallization of a magma, boiling (as in a hot spring), the cooling of a hot solution, and a chemical reaction between the fluid and the rocks through which it flows. There are a great variety of mineral deposits and several processes that lead to their formation. A conceptual model for the formation of mineral deposits, starting with the sourcing, transportation, concentration and deposition (Fig. 2.1).

Ore-forming processes are distinct offshoots of the main geological processes that form or modify rocks in the Earth's crust and land surface. There are four main geological processes responsible for the formation of rocks: (1) *igneous*, (2) *metamorphic*, (3) *sedimentary*, and (4) *surface weathering* (surficial). The first two processes are described as *exogenous* processes because they occur deep within the earth, while the last two are *endogenous because* they occur at or near the surface. The principal *concentrating* and *transporting* agents that are connected with the accumulation of economic minerals are:

(1) **Magmatic** - magma, and residual fluids,
(2) **Hydrothermal solution** - (magmatic water, metamorphic fluid, heated connate and meteoric water)
(3) **Sedimentary water -** river (running) water, seawater, connate water
(4) **Surficial** - meteoric water or groundwater
(5) **Metamorphic** - heat and fluids

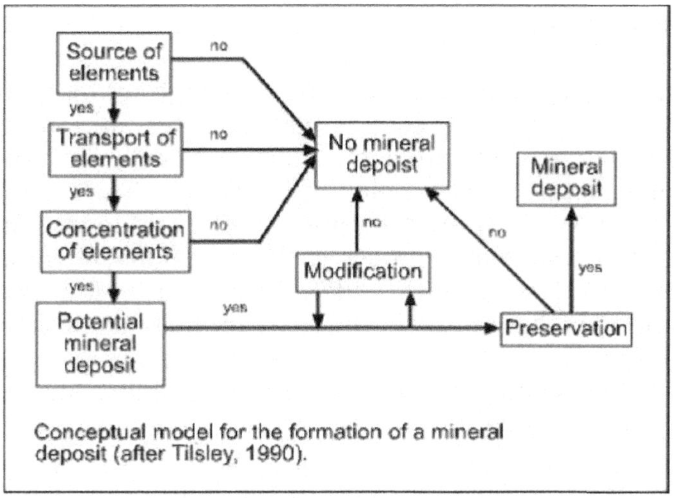

Fig. 2.1: Conceptual model for the formation of a mineral deposit

Magma is a hot molten rock that contains suspended mineral grains and dissolved gases that crystallizes as it cools. Economic minerals are concentrated through magmatic crystallization and segregation in which fractions of a magma can concentrate useful minerals. *Residual fluids* from magmas that are rich in metals and volatiles are able to form pegmatite deposits. *Hydrothermal solutions* are hot waters that serve as a concentrating, transporting, and depositing agent. They form the largest number of mineral deposits. The hot water may originate from various sources including crystallizing magma, heated connate water, seawater, meteoric water, and metamorphic water. On its own, *metamorphism* can concentrate or enrich economic minerals due to heat from regional metamorphism, or a combination of heat and fluids in contact metasomatic deposits.

Surficial processes such as chemical weathering involve the action of meteoric (rain) water in the chemical weathering and concentration of residual mineral deposits at the Earth's surface, such as bauxite and laterites, while downward percolating or migrating groundwater produce concentration of dispersed minerals near the oxidation-reduction boundary either by supergene enrichment or by groundwater infiltration. *Sedimentary wate*r either as flowing surface water or seawater are capable of concentrating useful metals or minerals and deposit or precipitate them in associated sediments.

The processes that lead to the deposition or precipitation of ore-transporting fluids are numerous but can be grouped or outlined as follows:

1. *Magmatic ore-forming processes*: - decrease in temperature, dissemination, gravitational settling, liquid immiscibility and fractional crystallization
2. *Hydrothermal ore-forming processe*s - decrease in temperature, boiling or degassing, changes in temperature and pressure, mixing of two solutions, reaction with wall rocks
3. *Sedimentary ore-forming processes* - chemical/biochemical precipitation, evaporation, clastic deposition, biogenesis and diagenesis.
4. *Metamorphic ore-forming* processes: - regional metamorphism (heat and pressure) and contact metasomatism (heat and fluids).
5. *Surficial ore-forming processes*: - weathering and supergene enrichment

2.2 Magmatic Processes

Magmas are hot molten rock materials that are liquid at high temperatures but as they migrate upwards from deep-seated sources, they cool as temperatures fall, and crystallize to form solid minerals. Economic minerals, just like other minerals in a rock, can crystallize from metal-rich magma, either as discrete crystals or accumulation of crystals. The sources of metals in magmas have been debated, but the most likely are the upper mantle and lower continental crust that may be enriched in such metals. Magmatic processes in a magma chamber can concentrate ore minerals by taking metals that were once widely dispersed at low levels in the magma and concentrating them in minerals that mostly accumulate in the body of the igneous rock or injected into surrounding rocks (Fig.2.2).

At times, parts of the magma that are rich in ore minerals, such as oxides or sulfides can concentrate largely as an "ore magma" which can separate and be injected into surrounding rocks by a process called "*filter pressing*" or magmatic injection. During magmatic differentiation or segregation, some dense ore minerals may sink to the bottom of the magma chamber due to gravitational settling, others may float to the top if they are lightweight minerals associated with volatiles, to be subsequently injected as low-temperature pegmatite deposits. Processes that concentrate and precipitate economic minerals in magmas include; dissemination, crystal settling, liquid immiscibility and fractional crystallization.

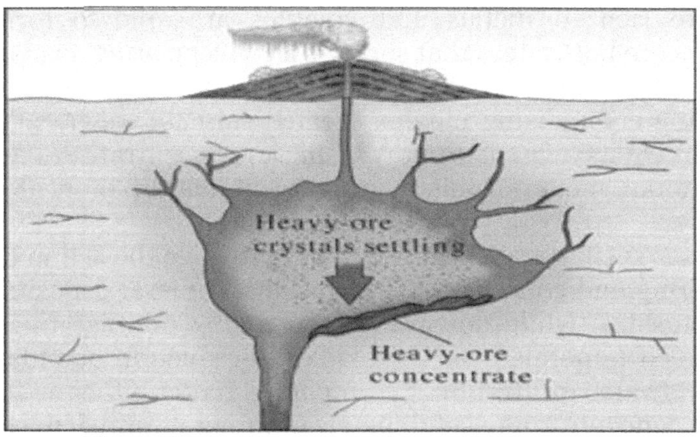

Fig. 2.2: Magma chamber and crystallization of silicate melt

2.2.1 Dissemination

This is the process of homogeneous scattering of ore minerals in a host rock during the early or late phase of magmatic crystallization Disseminations occur both in mafic and felsic magmas. This process may not produce deposits that are rich enough to be mineable at a profit, except if the dissemination occurs over a large area (e.g. porphyry copper) and at shallow depths, or if the deposit is of precious minerals such as diamond. Disseminated deposits produce most of the world's diamonds in kimberlites and niobium as columbite in alkaline granites and pyrochlore in carbonatites. The extensive disseminated copper and molybdenum sulfides in porphyry deposits that can extend for thousands of meters may be e.g., product of combined late-stage magmatic crystallization and hydrothermal precipitation.

2.2.2 Gravitational Settling

Some magma-related mineral deposits form layers within mafic-ultramafic rocks as a result of gravitational or crystal settling of early-formed, high density crystals to the bottom of a magma chamber. Crystal settling is a type of *magmatic segregation* - a process by which one or more minerals become locally concentrated (segregated) during the cooling and crystallization of a magma. Rocks and minerals formed by magmatic segregation are called magmatic cumulates. Mineral deposits that are magmatic cumulates are only found in mafic and ultramafic igneous rocks with low silica content. When there is excess silica in a magma, such as in granites, the crystals formed are forced to float rather than sink.

For example, one of the first minerals to form in mafic-ultramafic magma is chromite, the main ore of chromium. The dense chromite crystals settle to the bottom of the magma, producing almost pure layers of chromite, called *chromitite*. These ore-bearing layers often alternate with layers of the host rock. The world's principal deposits of chromite and platinum which are in the Bushveld Igneous Complex (BIC) in South Africa and the Great Dike of Zimbabwe were formed by this process.

Fig. 2.3: (a) Crystal settling (b) Chromite layers in the Bushveld Igneous Complex

2.2.3 Liquid Immiscibility

A different kind of magmatic segregation involves *liquid immiscibility*. This is a process by which a cooling magma will sometimes precipitate droplets of a second magma that has an entirely different composition. Such a magma may separate from the main body of magma to form a separate melt. Two liquids that will not mix are called *immiscible* (oil and water are an example). By this process a once homogeneous magma can separate into two or more magmas of different composition. When this happens, there are usually two liquids, one silicate-rich and the other, typically much smaller in volume, can be rich in metal *oxides*, *sulfides* or *carbonates*. The sulfides or oxides may separate and sink below the silicate-rich part of the intrusion (Fig. 2.4) or be injected into surrounding rocks by a process known as "filter pressing" to form massive ores. Examples are found in mafic-ultramafic rocks such as the massive nickel sulfide and platinum deposits in the Bushveld Igneous Complex, South Africa, and the titano-magnetite deposits in anorthosites (e.g. Allard Lake, Quebec, Canada).

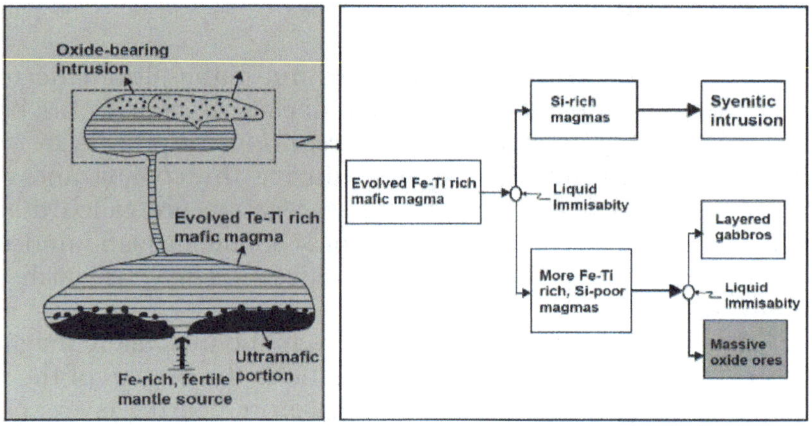

Fig. 2.4: Schematic diagram of 2-stage immiscible Fe-Ti melts

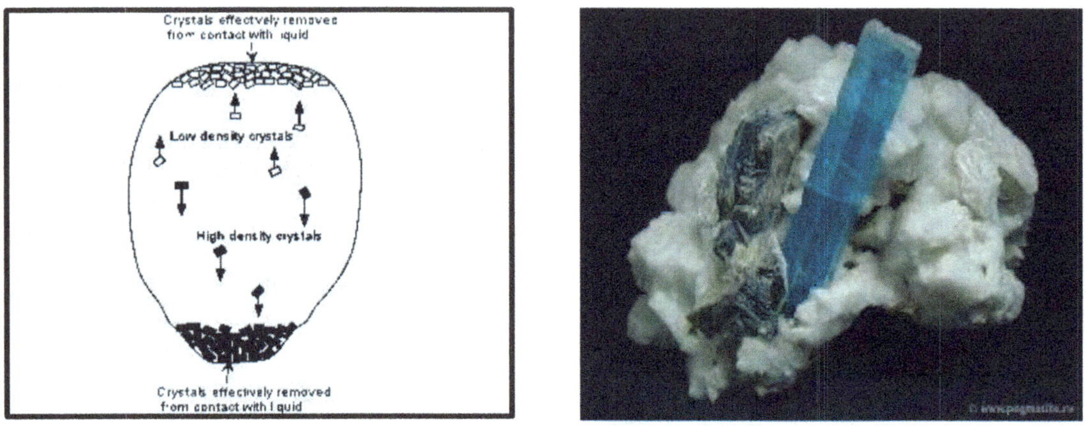

Fig. 2.5: (a) Fractional crystallization in a magma (b) Pegmatite with blue gemstone crystal

2.2.4 Fractional Crystallization

Fractional crystallization is one of the main processes of magmatic differentiation. This is the removal of early formed crystals as the magma crystallizes which changes the composition of the magma and results in the enrichment of certain elements and volatiles in the residual fluids (Fig. 2.5a). Magmas normally contain water as vapor that may reach up to 10% of the magma's volume. The volatiles and enriched elements become a fraction that may crystallize within the igneous body or are injected into surrounding rocks to form new minerals some of which are of economic value. Examples are pegmatites and alaskites (feldspar rich rock) that originate from granitic magmas. Pegmatites are found in most continents and are a major source of the world's lithium, beryllium, cesium, niobium, and tantalum deposits and gemstones such as tourmalines, emerald, aquamarine and garnets. (Fig. 2.5b).

2.3 Hydrothermal Processes

Hydrothermal processes precipitate mineral deposits as a result of the actions of hydrothermal solutions and their reaction with surrounding rocks. Hydrothermal solutions are the hot aqueous fluids (hot water) that serve as a very effective medium for the transportation and enrichment of certain elements which are subsequently precipitated as economic minerals. There are numerous reasons why hydrothermal solutions will deposit their load of minerals, but these will be discussed in detail in Chapter 4. The sources of the hot solutions range from magmatic to meteoric. Mineral deposits formed by hydrothermal processes are known as hydrothermal deposits. They are the most numerous of all classes of deposits. Hydrothermal solutions are not ordinary hot water but Na-Ca chloride brines that are effective solvents of many sulfide and oxide ore minerals, and are capable of dissolving and transporting native metals such as gold and silver. Their temperatures range from (less than 50°C to 650°C depending on their source.

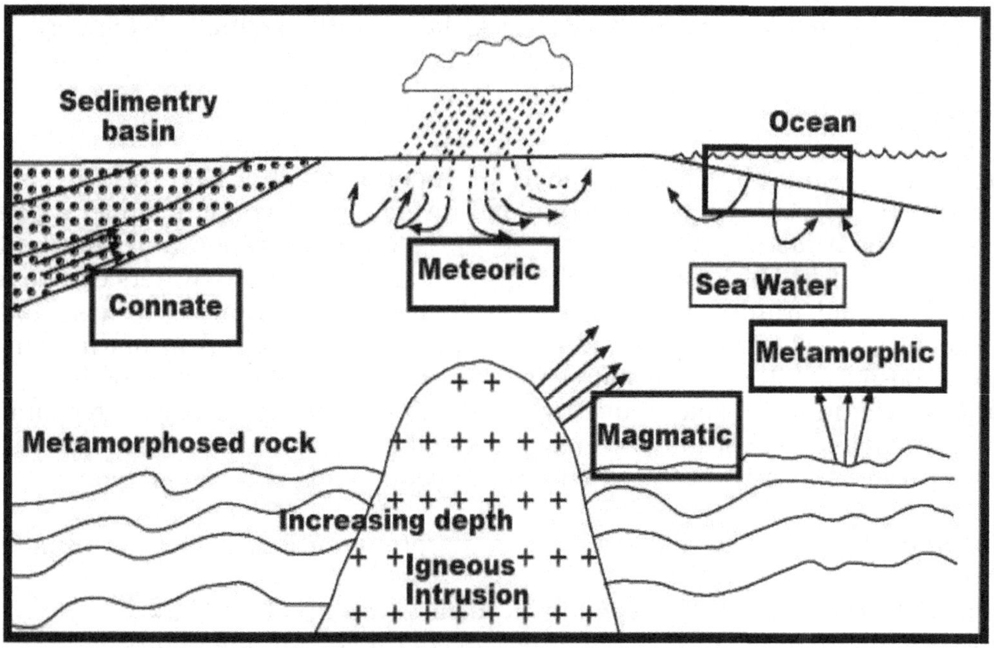

Fig 2.6: Sources of hydrothermal solutions

Hydrothermal solutions can be derived from the following five sources: magmatic water, connate water, sea water, meteoric/groundwater and metamorphic water. Figure 2.6 shows the various sources of hydrothermal fluids. Hydrothermal solutions can originate from the following five sources: magmatic water, connate water, sea water, meteoric/groundwater and metamorphic water (Fig. 2.6) Hydrothermal solutions are capable of transporting metals and ore elements that are derived from several sources, including magmatic fluids, heated connate brines, leaching of metals from crustal rocks and oceanic volcanics; dissolution of low-temperature minerals such as carbonates and mineral salts, and the breakdown of minerals due to elevated temperatures during metamorphism. The main reason why metal-bearing hydrothermal solutions precipitate economic minerals is due to the loss of physical and/or chemical stability. Hydrothermal deposits are of several types but the most important are vein deposits, porphyry deposits, skarn deposits, volcanogenic massive sulfide deposits, sedimentary exhalative deposits, and Mississippi-Valley type deposits (Fig. 2.7).

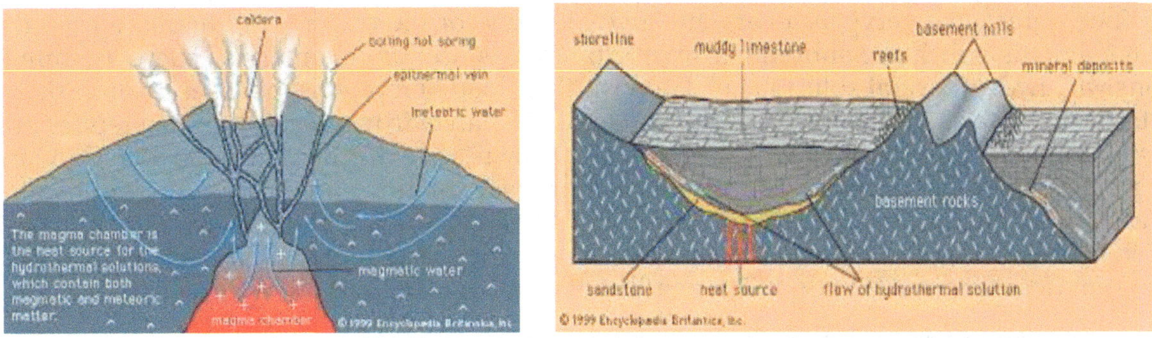

Fig. 2.7: Hydrothermal deposits formed in association with magmatic & meteoric waters

2.4 Sedimentary Processes

Sedimentary processes take place at relatively low temperatures at or near the Earth's surface and contribute to the transportation, concentration and formation of economic mineral deposits. Such processes include chemical precipitation from sea or lake water, evaporation of seawater, alluvial deposition and diagenesis. Heavy metals held in solution in sea or lake waters can be precipitated or chemically deposited from solution whenever their concentrations exceed solubility ranges. Also, some biochemical reactions can lead to precipitation of minerals in marine environments. Sedimentary deposits are the main sources of the world's metallic ores of iron, manganese, copper, titanium, uranium and gold, industrial minerals and rocks (e.g. barite, gypsum, phosphate rock and limestone), and energy minerals (e.g. coal and tar sand). Mineral deposits are formed in sedimentary environments by processes that can be grouped into three main types:

a) *physical* accumulation during seawater evaporation and sediment entrainment, transport and deposition;
b) *chemical/biochemical* precipitation from surface water bodies; and
c) *biogenic* accumulation of the skeletal remains of organisms and plants.

2.4.1 Clastic Deposition

The *physical* processes of erosion, transportation by rivers and streams and deposition of clastic materials including sand, silt, clay and resistant (heavy) minerals can lead to accumulation of valuable minerals. The process starts with the physical and/or chemical disintegration of rocks and minerals which can lead to removal of the heavy minerals by erosion along slopes (eluvial) that ultimately reach the streams and rivers. Streams and coastal currents are important agents for separating and sorting pre-existing precious or heavy minerals. Alluvial mineral deposits are formed in rivers and streams by hydraulic concentration of heavy, durable minerals such as gold, tin, platinum, and diamonds. There are certain sections of streams where heavy minerals accumulate such as the inner bends of meanders (Fig. 2.8). Along the coastline, coastal sea-currents can concentrate dense minerals along beaches. Alluvial and beach placer deposits are formed by this process, and major examples include gold, cassiterite and gemstones along river beds and magnetite (black sand), ilmenite, rutile and zircon found along beaches. They are mined extensively in several countries including the Philippines, South Africa and Indonesia.

2.4.2 Evaporation

Evaporation is a *physical* process in which water or aqueous solutions vaporize into the atmosphere due to the effect of heat which may lead to the concentration and precipitation of useful minerals. In the surface environment, a large body of water trapped in a shallow sea or lake may dry out during warm periods to produce extensive deposits of evaporites. There are two types of evaporite deposits: *marine* evaporites formed from ocean or sea water, and *non-marine* evaporites formed from smaller standing bodies of water, such as lakes. Large mineral deposits of halite (NaCl), gypsum ($CaSO_4$), borax, and potash (KCl) are formed by this process.

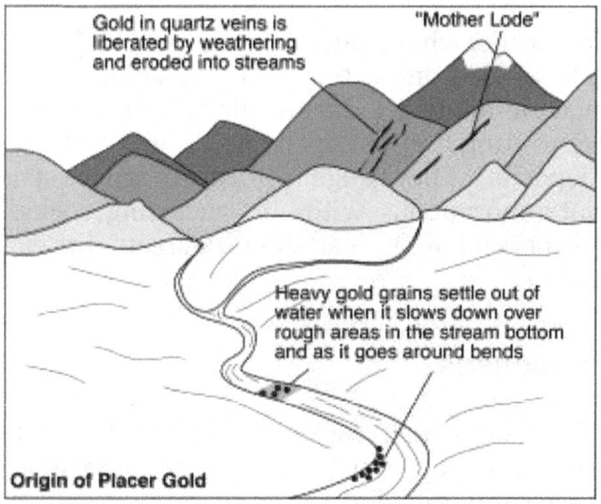

Figure 2.8: Formation of placer deposits

2.4.3 Chemical/Biochemical Precipitation

Chemical and/or biochemical processes can lead to the precipitation of metals and other valuable minerals held in solution in standing bodies of water due to reactions, changes in physical chemical and biochemical conditions or when their concentrations exceed solubility ranges. These metals or minerals are chemical precipitates that accumulate as sedimentary mineral deposits on the lake bottom or seafloor (Fig. 2.9). The most noteworthy examples are the iron formations, and sedimentary copper and manganese deposits.

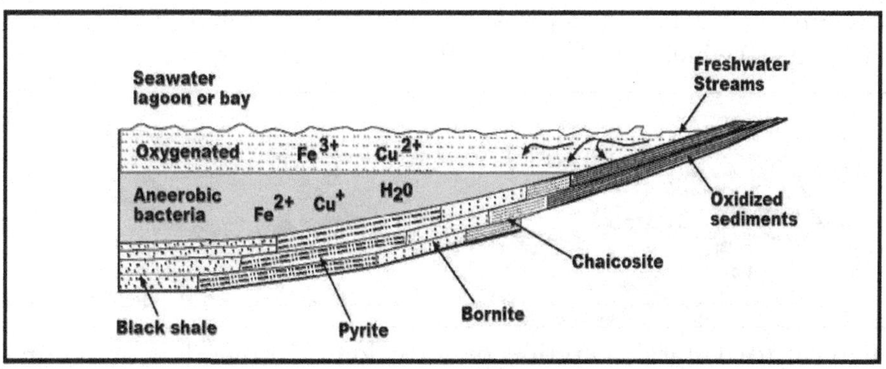

Figure 2.9: Chemical precipitation of metals

2.4.4 Biogenic Processes

Biogenic processes can lead to the accumulation of the remains of plants and animal skeletons in near-shore to deep marine environments. Rocks that form from such processes may be of economic value. Examples include rocks formed from calcareous and siliceous skeletal remains of marine organisms such as limestones, diatomite and phosphate, and coal and lignite from plant remains.

2.4.5 Diagenesis

Diagenesis is the totality of all the chemical, physical and biological changes that occur after sediments are initially deposited, compacted and consolidated (excluding metamorphism). Mineral deposits are formed or modified during diagenesis by alteration, remobilization and leaching of metals and concentrating them in open spaces within the sediments. Such deposits are called “diagenetic mineral deposits". They range from stratiform to stratabound, and syngenetic to epigenetic but are still contemporaneous with the enclosing rocks. Examples include copper deposits in shales (e.g. Kuperschiefer), lead-zinc deposits in carbonates and uranium in sandstones (e.g. Niger deposits)

2.5. Metamorphic Processes

Metamorphism is the process by which rocks are changed due to the effects of increased temperature and/or pressure, and addition of fluids. Metamorphism can be regional or local (contact), and both can produce mineral deposits.

Regional metamorphism may cause the formation of new minerals, recrystallization of existing minerals, and mobilization of pre-existing fluids in rocks capable of leaching metals and concentrating them in fissures and other permeable zones as mineral deposits. Economic minerals can therefore be formed during metamorphism in three ways: (1) crystallization of new minerals, such as graphite or garnets, known as truly “*metamorphic*” deposits; (2) recrystallization of existing minerals which may produce increased grain size (coarser) or transformation of existing minerals of the same composition designated as *metamorphosed* deposits (e.g. hematite ($Fe2O3$) can recrystallize to coarse grained specularite ($xFe2O3$) or change to magnetite ($Fe3O4$); and (3) remobilization of ore bearing metamorphic fluids, (e.g. manganese or gold) and concentrating them into fracture zones. This type of deposit has been named as “*metamorphogenic*” by Walter H. Pohl (2011). Lateral secretion of metals from country rocks into fault zones and other structures can create rare mineral deposits during dynamic metamorphism (Fig. 2.10) which may belong to the metamorphorgenic group of deposits.

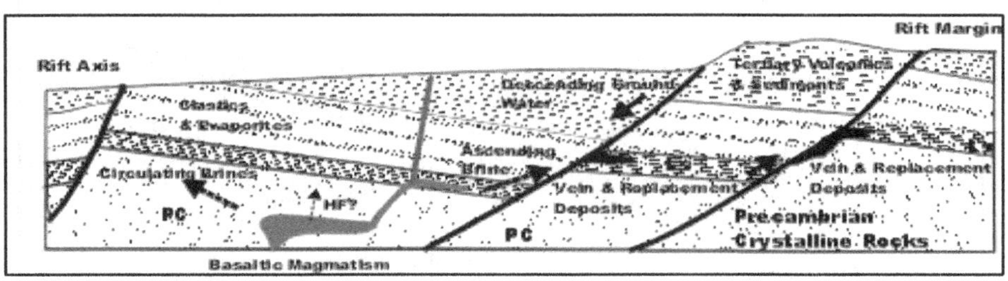

Figure 2.10: Lateral secretion or diffusion of minerals from country rocks into faults and other structures during metamorphism

Contact metamorphism involving the emplacement of igneous rocks into sedimentary sequences (particularly limestones), may produce chemical reactions that lead to the formation of mineral deposits. These chemical reactions involving fluids emanating from silicate melts (granitoid intrusion) often lead to the formation of calcium-silicate minerals (calc-silicate) minerals such as wollastonite, and zeolites. What is more significant is when the process includes the addition of metal-rich fluids (*metasomatism*) from the intrusive rocks which may lead to the extensive replacement of surrounding rocks and deposition of economic minerals. Mineral deposits of this type are called *"skarn"* or contact metasomatic deposits. Examples are copper and tungsten deposits in western Canada and the United States.

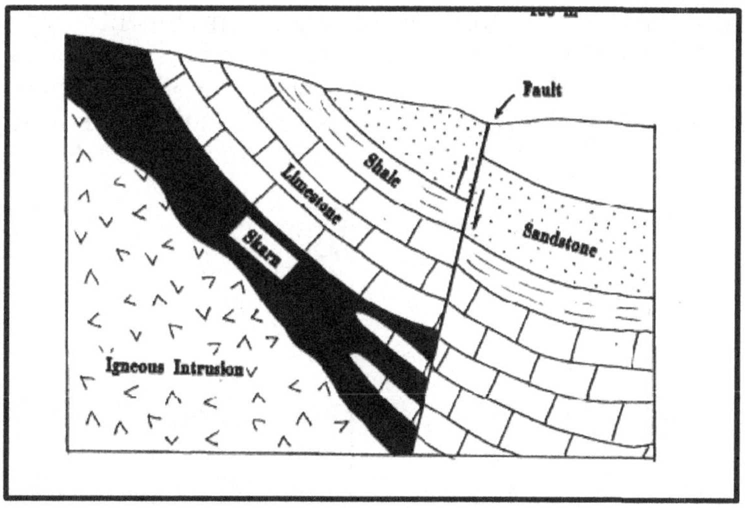

Fig. 2.11: Skarn deposit replacement of limestone

2.6. Surficial Processes

Surficial processes are those that take place at or near the Earth's surface involving the actions of rain (meteoric) water and downward percolating groundwater during weathering and related processes. There are two main types of weathering; *chemical* and *mechanical* weathering, and both types may result in the concentration of useful economic minerals.

Chemical weathering involves the breakdown of rocks through the process of oxidation and hydrolysis. This process leads to the decomposition of minerals accompanied by selective leaching and concentration of useful elements that can form secondary minerals of potential economic interest. Such concentrations are called "residual" deposits, and examples include bauxite deposits, and lateritic iron, nickel and manganese deposits. They are most common in tropical and subtropical regions where high temperatures combined with heavy precipitation often leads to deeply weathered profiles. Most of the world's supply of bauxite (source of aluminum) derives from residual blanket deposits. Lateritic nickel ores formed by intensive tropical weathering of ultramafic rocks containing approximately 0.3% nickel occur in Australia and Papua New Guinea. The initial nickel content has been strongly enriched by selective leaching in the course of lateritic weathering that made the deposits of economic value.

Mechanical weathering which involves the physical (rather than chemical) breakdown of rocks can release and concentrate minerals that possess high specific gravity. This process allows heavy minerals such as, zircon, rutile, chromite, monazite, diamond and gold nuggets to accumulate by erosion and gravitational settling. The physical breakdown of rocks and minerals in temperate climates where there is low precipitation and cool weather is more conducive to mechanical weathering and formation of eluvial placer deposits. This is in contrast to tropical environments where chemical decay of rocks is pervasive.

Supergene processes are a remarkable form of intense chemical weathering that concentrate minerals by secondary enrichment, through the action of descending surface waters. They usually involve hydrolysis and oxidation of sulfide and/or oxide ores and precipitation at shallow depths below the surface usually close to the water table interface (Fig. 2.12a). Supergene enrichment can apply to many transition metals, but copper (Cu) is the outstanding example of ores formed by this process.

Meteoric water infiltration is a form of supergene enrichment in which highly oxygenated groundwaters are capable of moving through sedimentary strata leaching elements that are later precipitated due to changes in the Eh-pH conditions usually along a defined reduction front (Fig. 2.12b). Examples of elements subject to this type of leaching are uranium, vanadium and molybdenum. The mineral accumulations formed by this process are called "infiltration deposits".

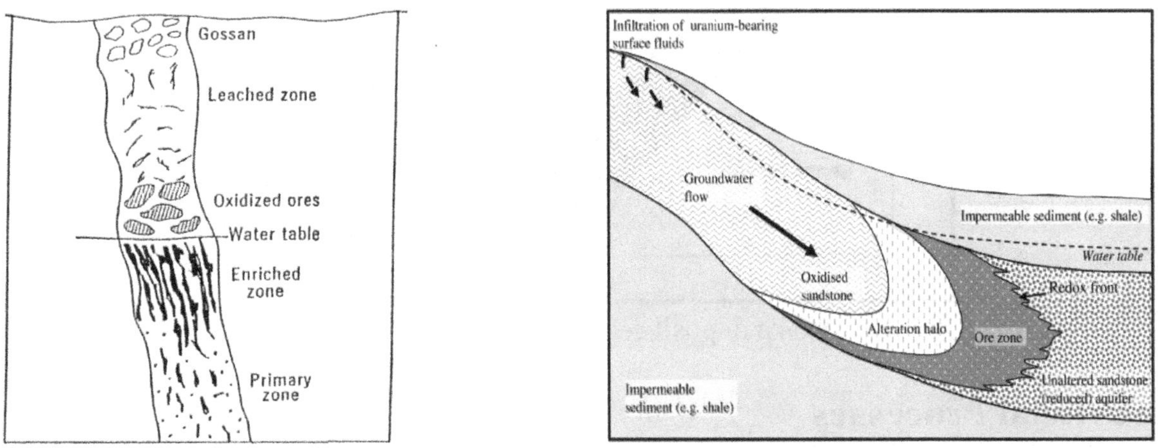

Figure 2.12:(a) Zones of Supergene Enrichment (b) Meteoric Water Infiltration

Review Questions

1. Discuss the major processes of ore formation and explain their connection to the formation of their host rocks.
2. What are the possible sources of metals in magmas and how can they be concentrated during magmatic crystallization?
3. Briefly describe the physical and chemical/biochemical processes in sedimentary environments that lead to the concentration and precipitation of mineral deposits.
4. Explain the major processes that lead to the formation of metamorphic, metamorphosed and contact metasomatic deposits, citing examples.
5. What are the differences between lateritic weathering and supergene enrichment? List two examples of deposits formed by each process.

Chapter 3: Magmatic Mineral Deposits

3.1 Introduction

Magmatic mineral deposits (or orthomagmatic deposits) are the accumulations of economic minerals within igneous rocks. They are usually formed deep within the Earth's crust during crystallization of plutonic rocks, and rarely formed at the surface in volcanic rocks. Magmatic mineral deposits vary in shape and size, from irregularly shaped, flat-lying (stratiform) to lenticular and pod-like form (or podiform), and from tabular to tubular in morphology. The host rocks range from ultramafic to ultrafelsic in composition, and the deposits can form massive ores, layers and lenses in some cases, and disseminations of rare minerals in others. This group of deposits are the biggest and best-known sources of economic minerals, including diamond, copper, niobium, tin, platinum, titanium and chromium. Examples of magmatic deposits are: diamond in kimberlites, chromite, nickel-copper sulfides, iron-titanium oxides, and platinum group metals (PGM) in mafic/ultramafic rocks, while tin, niobium, uranium and copper are found in felsic rocks, and niobium and rare earths in carbonatites.

The close association between mineral deposits and igneous host rocks is well known and described in economic geology. The strong affinity of some metals and rare elements with a particular type of magma or igneous host rock is known as the "*ore-rock association*". This relationship is believed to be genetic, and attributable to the fact that certain metals tend to originate from a particular type of magma because of their chemical composition, that is ultramafic, mafic, felsic, ultra-felsic, alkaline and peralkaline types. The following are well-known associations between mineral deposits and types of igneous host rock.

1) Deposits related to *ultramafic and mafic* intrusive rocks. (Cr, Pt, Ni-Cu and Ti-Fe).
2) Deposits related to mafic volcanic rocks (Cu, Ni)
3) Deposits related to *felsic granitic* rocks (Cu, Sn, W)
4) Deposits related to *ultra-felsic* pegmatites (Sn, Ta, Nb, U)
5) Deposits related to *alkaline* granites *and* carbonatites (REE, Nb, U)

3.2 Types of Magmatic Mineral Deposits

Magmatic mineral deposits can be classified into four groups based on the process of formation and morphology.

1) Disseminated deposits (Ni, Cr, Nb, REE
2) Layered (Stratiform) deposits (Cr, PGE, Fe, Ti, V)

3) Massive Sulfide deposits (Ni, Cu, PGE) - (volcanic and plutonic subtypes)
4) Pegmatite deposits (Sn, Nb, Ta, Li, U)

Figure 3.1 is a schematic diagram illustrating the various types of magmatic mineral deposits. On top of the diagram is the basaltic lava (komatite) hosted massive sulfide Ni-Cu deposits, below which is a funnel-shaped intrusive complex with layered deposits of Cr, PGE, Ti-Fe which pass downwards into varieties of massive Ni, Cu sulfides with PGE in the mafic intrusive body and feeder dyke. Pegmatite deposits of Sn, Ta, Li, Be and alkaline intrusive rocks with disseminated Sn, REE and Nb are shown at the bottom right corner of the diagram.

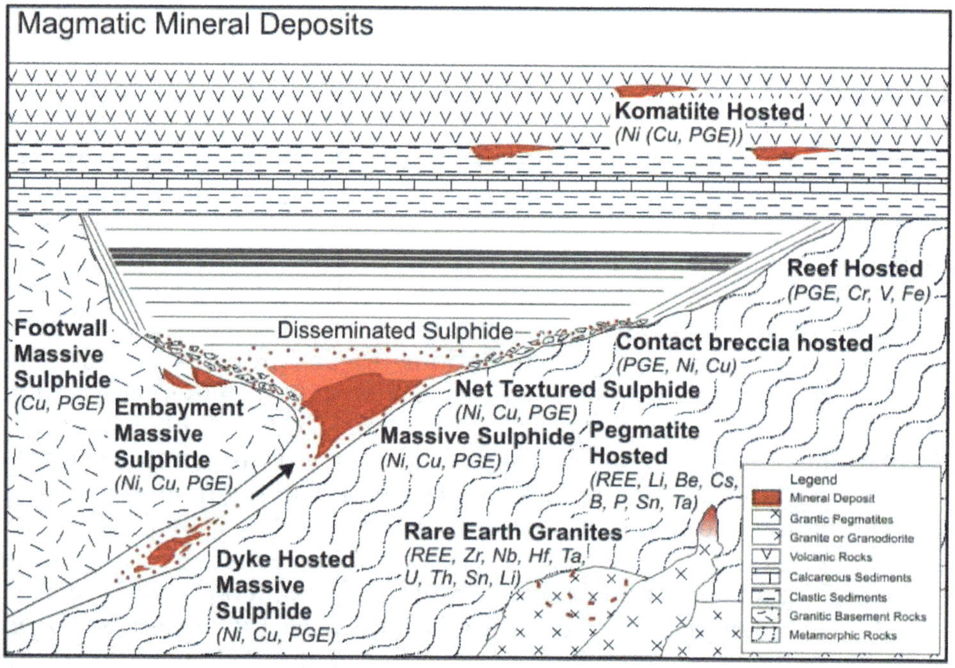

Fig. 3.1: Schematic diagram of magmatic mineral deposits

3.2.1 Disseminated Deposits

Disseminated deposits are characterized by the random or uniform scattering of ore minerals in igneous host rocks. The disseminated ore usually contains small particles of valuable minerals, sometimes invisible, that are spread throughout the host rock. These deposits are found in rocks derived from both mafic and felsic magmas. Depending on the type of magma, the dissemination may be early magmatic (e.g. chromite in mafic rocks) or late magmatic as for columbite in alkaline rocks. Examples of disseminated deposits are diamonds in kimberlites, cassiterite (Sn) and columbite (Nb) in alkaline granites, pyrochlore (Nb) in carbonatites and disseminated copper sulfides in granites of porphyry deposits For many of these disseminated deposits, ore grades are usually low, but the concentrations may be adequate to make the mineral deposit economic if the ore minerals are of high value, such as diamonds, or dispersed over a large area (large tonnage) to be mined profitably, or enriched by secondary processes such as supergene enrichment of Nb deposits in the Honore carbonatite, Brazil.

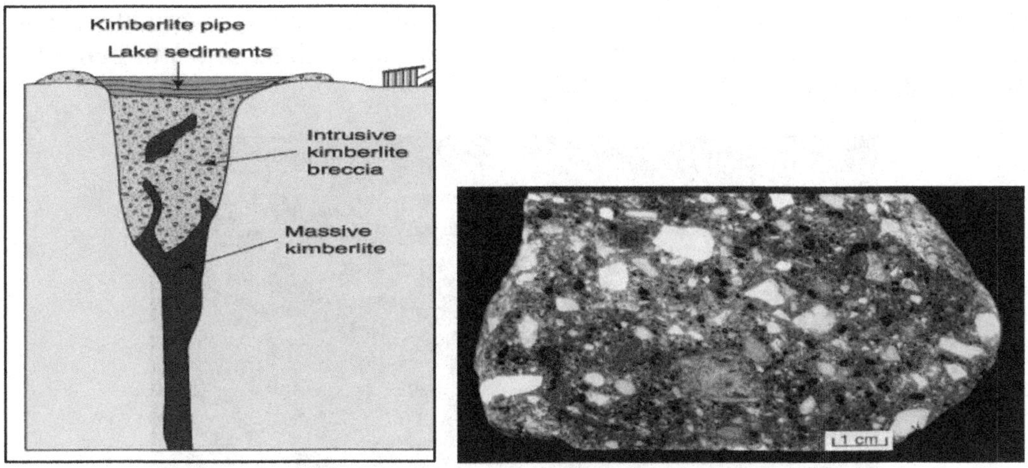

Fig. 3.2: Kimberlite pipe and kimberlite breccia from 'pipe'

Disseminated deposits produce most of the world's diamonds from kimberlites which are unusual potassium-rich, low-silica volcanic rocks of ultramafic origin that rose from the mantle by explosive activity. Diamond occurs as discrete crystals or fragments scattered randomly within the kimberlite that rarely shows any discernible alteration (Fig. 3.2). Diamond-bearing kimberlites are known to occur within ancient cratonic shields of the world and are found in several African countries including South Africa, Botswana, Angola, Democratic Republic of Congo, Tanzania, Lesotho, Ghana, Sierra Leone, Guinea and Liberia. Other world leading producers are Russia, Canada and Australia.

3.2.2. Stratiform (Layered) Cumulate Deposits

These are ore deposits that occur in layered mafic and ultramafic intrusions as cumulate deposits in massive sheets, bands, and lenticular bodies that are conformable with the primary layering of the host rocks (stratiform). This type of mineral deposits is only found within mafic and ultramafic rocks due to the control exerted by low silica on the viscosity of a magma. Magmatic cumulate deposits are derived from the accumulations of crystals of metallic oxides that formed by crystal settling or by immiscible sulfide or oxide liquids that concentrated during the crystallization of ultramafic-mafic magma to form layers of massive ores. Metals that accumulated in these deposits are mainly chromium, platinum group metals, nickel-copper and titanium-iron. They contain most of the world's deposits of platinum-group metals and chromium. Classical examples of stratiform magmatic deposits are found as chromite and platinum-rich in the Bushveld Igneous Complex (BIC) of South Africa, the Great Dyke of Zimbabwe and Stillwater complex, Montana (Fig. 3.3).

The vast Bushveld Igneous Complex (BIC) is a large funnel-shaped layered mafic to ultramafic igneous body with some layers consisting of 90% chromite associated with platinum-group elements (PGEs) platinum, palladium, rhodium, iridium, osmium and ruthenium—along with vast quantities of titanium and vanadium. The dominant lithologies within the BIC vary from ultramafic peridotite, harzburgite, and bronzitite in the lower sections of the unit, to mafic norite, anorthosite, and gabbro towards the top. The complex contains the world's largest reserves of chromium and PGEs. The chromitite seams are cyclic in nature as well as laterally contiguous throughout the entire intrusion. Gangue minerals include olivine, pyroxenes, plagioclase, and sulfides. Platinum is found in the chromite layers but not in economic quantity. The economic concentrations are localized in stratiform PGE-enriched mineralized zones in mafic to ultramafic layered intrusions known as "reefs" (Fig. 3.4).

Fig. 3.3: Chromite layers in the Bushveld Igneous Complex, South Africa

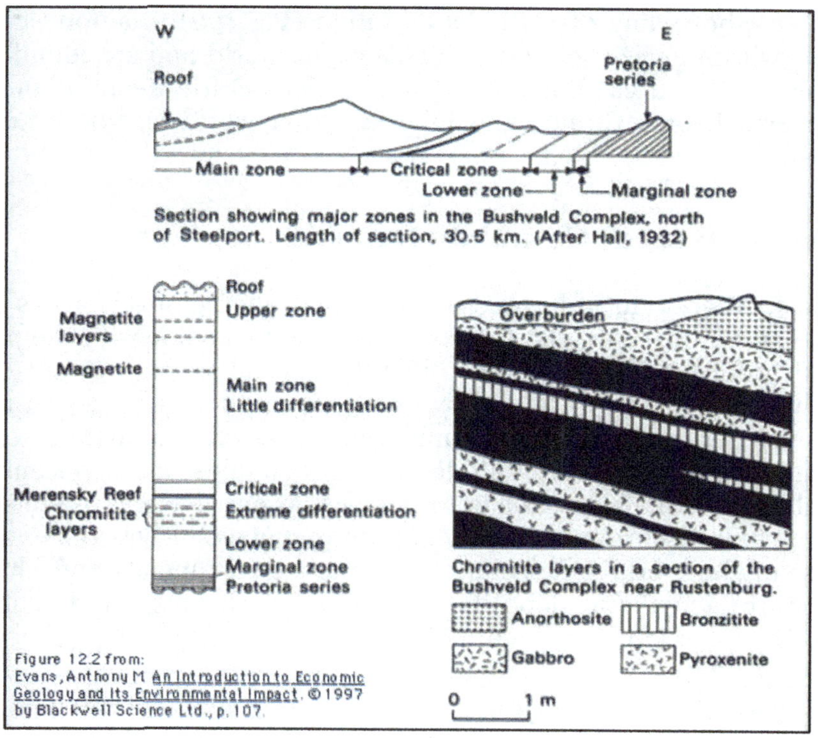

Fig. 3.4: Vertical section through Platinum-Chromite deposit

3.2.3 Massive Sulfide Deposits

A group of nickel-copper (Ni-Cu) sulfide deposits occur predominantly within mafic/ultramafic intrusions, and rarely in ancient (Archean) basaltic lavas known as *komatiites*. They are sometimes closely associated with platinum-rich and chromite layers. The massive sulfides are also stratiform and occur as sulfide-rich layers intercalated within or near the base of the plutonic body and lava flows. Prominent examples are the Ni-Cu deposits in the Sudbury Igneous Complex in Ontario, Canada (Fig. 3.5), and in komatite lavas in Kambalda, western Australia (Fig. 3.6),

Most of the very large Ni-Cu sulfide deposits around the world belong to this group. They are found within small mafic intrusions where they accumulate at or near the base of the intrusions. The deposits are stratiform, blanket-shaped bodies consisting of massive sulfides, in layered intrusions that are commonly gabbro-norite or anorthosite in composition. The sulfide ore minerals are pentlandite (Ni), chalcopyrite (Cu), and pyrrhotite/pyrite (Fe), and may be associated with platinum (PGE) and minor chromite. The ore grade is often low, but adequate for economic extraction of Ni and Cu, sometimes as supplementary metals to Cr and platinum. These mineral deposits are believed to have been concentrated by liquid immiscibility involving the segregation of sulfide-rich melts near the base or edges of the mafic magma chamber.

The massive Ni-Cu deposits that occur in the Sudbury Igneous Complex in Ontario, Canada are a classic example of this type of mineralization. The complex is a large, oval 65 by 35 km basin-shaped mafic intrusion about 1.85 billion years old, and believed to be of meteorite impact origin (Fig.3.5). The individual ore bodies are moderately dipping, lenticular, or sheet-like bodies of disseminated, massive, or breccia ore or sulfide stringer zones. In some areas, the ore bodies coincide with steep faults along the norite contact.

The Kambalda deposit in Western Australia occurs within komatite, a unique ultramafic volcanic rock containing low silica (40-45%) and potassium, but extremely high magnesium content (>18%). Komatiites are very rare and restricted mainly to the Precambrian. Mineralization of Ni and Cu ores occur within the lavas concordant with the host rocks mainly as massive sulfides that are composed of >90% Ni-Cu-Fe sulfides while the matrix sulfides are composed of 30-50% sulfides interstitial to olivine. The dominant ore minerals are the sulfides of Ni, and Cu (pentlandite and chalcopyrite with pyrite) and occasional PGEs. The origin of the Kambalda ores is attributed to crystallization from an immiscible sulfide liquid that had become segregated by liquid immiscibility from a komatiite lava (Fig. 3.6).

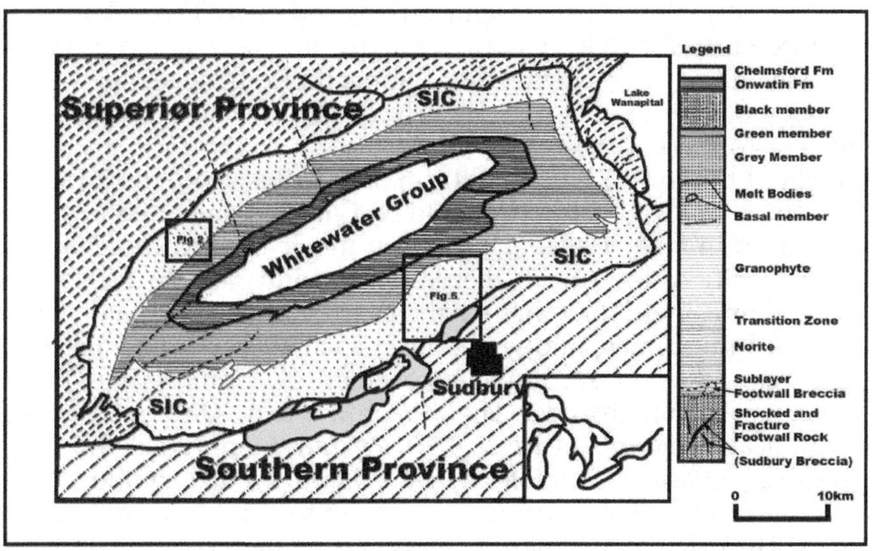

Figure 3.5 General geology of Sudbury Igneous Complex, Ontario, Canada

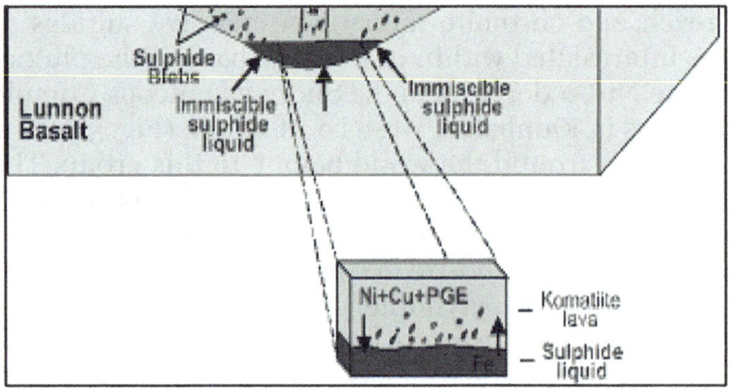

Figure 3.6: (a)Massive sulfide deposit in komatite lava (b) Ni-Cu-PGE Ore in mafic rock

3.2.4 Pegmatite Deposits

Pegmatites are very coarse-grained igneous bodies formed from residual volatile-rich fractions of granitic magma by the process of crystallization fractionation or magmatic differentiation. As a granitic intrusion reaches the final stages of cooling, the residual water-rich, low-temperature fraction concentrates at the top along with volatiles and other elements (with large ionic radius or small ionic charge) that could not enter the silicate structure of minerals. This residue may be injected into fractures within or surrounding the igneous intrusion to form pegmatites that may contain ore minerals. The most common silicate minerals are K-feldspar, muscovite and quartz. Other minerals are tourmaline, amazonite and accessory gemstones. Mineral zoning of complex pegmatites is common and it involves different mineral assemblages within each zone.

Pegmatite deposits generally occur as dykes that sometimes extend for several hundreds of meters usually in proximity with granites (Fig. 3.7a). They occur as individuals or swarms that may be localized within or at the borders of the granitic intrusive rocks from which they emanated. Pegmatites are often rich in elements of economic and strategic importance, such as Lithium (Li), Cesium (Cs), Tantalum (Ta), Niobium (Nb), Boron (B), Beryllium (Be), Tin (Sn), Uranium (U) and the Rare Earth Elements (REE). Some pegmatites contain useful semi-precious and precious gemstones including aquamarine, tourmaline, topaz, fluorite, apatite and corundum. Pegmatites are found in most continents, particularly in Africa where they are sources of tin, tantalum, lithium and gemstones (Fig. 3.7b)

Much of the world's lithium (Li) is derived from pegmatites, from the minerals spodumene and lepidolite; such as those at King's Mountain, North Carolina, and Bikita in Zimbabwe. The great Tanco pegmatite in Manitoba, Canada, produces much of the world's cesium from the mineral pollucite. The majority of the world's beryllium is sourced from non-gem quality beryl within pegmatite. Tantalum, niobium, rare-earth elements are sourced from pegmatites Feldspars, which are used in ceramics, and gemstones, such as beryl, aquamarine, tourmaline, ruby, and emerald, are very rare and are found in pegmatites. Uranium has been exploited in pegmatites where it occurs as pitchblende. The Bancroft Field, Ontario, Canada is a major producer of U, with significant amounts of REE, Y, Ti, Zr and Be minerals also present. Another huge deposit is the Rossing Deposit in Namibia, which is large tonnage, but low grade. The Older Granite pegmatites in Nigeria are major sources of tin, niobium, tantalum and a suite of industrial minerals and gemstones. The origin of pegmatite deposits is related to residual melts produced by late-stage fractional crystallization of volatile-rich granitic melts. Several unusual elements and rare earths may accumulate under conditions of extreme fractionation resulting from extended crystallization of quartz and feldspars.

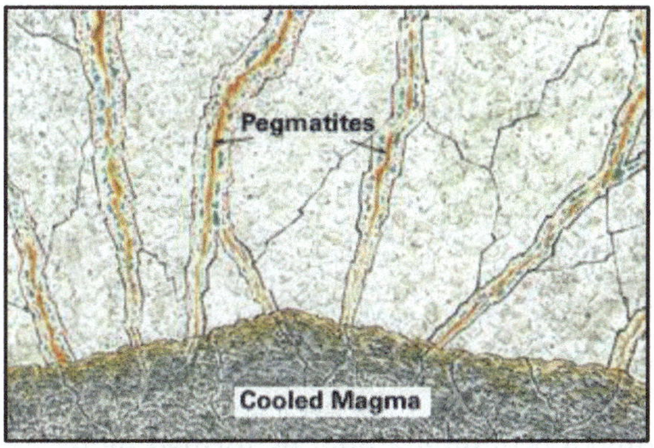

Fig. 3.7: (a) Schematic diagram of pegmatite dikes. (b) Massive pegmatites in Tanzania

Review Questions

1. List the main types of magmatic deposits and explain their inter-relationships.
2. Describe the concept of "ore-rock association". List three types of igneous rocks and their associated mineral deposits.
3. Briefly describe the mode of occurrence of platinum and chromite deposits. Expatiate on their origin.
4. What are the characteristics of massive sulfide deposits? List their metallic minerals and explain the origin of the deposits.
5. Describe the mode of occurrence and mineralogical composition of pegmatites.

Chapter 4: Hydrothermal Mineral Deposits

4.1 Introduction

Hydrothermal mineral deposits are concentrations of economic minerals formed by precipitation from hot mineralizing solutions into open spaces or fissures in rocks. Hydrothermal deposits are formed at a wide range of depths, from near the surface to up to 10km in depth. The temperature of formation ranges from about 500°C to as low as 50°C. These varying conditions of temperature and pressure can affect the conditions of ore formation as well as the stable mineral assemblage. At higher temperatures, certain ore minerals are more stable such as cassiterite (Sn), and the same situation applies at low temperatures where minerals such as silver (Ag) sulfides and cinnabar (Hg) are more stable.

Hydrothermal deposits are mainly epigenetic while a few may appear syngenetic or diagenetic. They are formed in a variety of host rocks, but there may be obvious preferences for particular types of rock. For example, certain lead-zinc deposits have preferences for carbonate host rocks while hydrothermal copper deposits may occur preferentially in fine clastic sediments such as shales and siltstones. The mode of occurrence includes veins and veinlets, stockworks and tabular bodies. Structural control, both at regional or local levels is a major factor in the localization of hydrothermal deposits. Hydrothermal fluids usually originate from various sources, including water dissolved in magma (magmatic), deep-seated water systems, heated groundwater (meteoric), sea water and connate water that circulate within sediments. Metamorphic processes may heat up and remobilize connate or ground water to become potent mineralizing fluids that could be channeled into fissures.

Hydrothermal deposits are leading sources of metallic ores worldwide. They are of enormous significance in the production of many important metals such as copper, lead, zinc, gold, silver, and radioactive metals. In addition, these deposits serve as the sources of industrial minerals such as asbestos, fluorspar, barite, graphite, and several semi-precious stones (amethyst, tourmaline, topaz, and beryl).

4.2 Localization_of Hydrothermal Deposits

Hydrothermal deposits are localized within fractures, fissures and other available openings in host rocks, usually by precipitation of minerals from hot solutions due to changes in temperature, pressure or reaction with wall rock. There are two main processes of deposition; *cavity filling* and/or *replacement* of host rocks. The two processes may occur simultaneously, the filling of an opening by precipitation may accompany the replacement of the walls of the opening. Transport by hydrothermal solutions usually requires a salt or other soluble species that can form metal-bearing complexes. These metal-bearing complexes which occur as hydroxides, chlorides etc., facilitate transport of metals within aqueous solutions generally by processes similar to chelation. (Chelation is a particular way in which ions and molecules bind metal ions).

When a hydrothermal solution moves slowly upward, as with heated groundwater percolating through an aquifer, the solution cools very slowly. If dissolved minerals were precipitated from such a slow-moving solution, they would be spread over a large volume of rock and may not be sufficiently concentrated to form an ore. But when a solution flows rapidly, as in an open fracture, or through a mass of shattered rocks, or through a layer of porous rock where flow is less restrained, cooling can be sudden and can occur over short distances. This usually results in rapid precipitation and a concentrated mineral deposit is formed. The particular minerals precipitated in these deposits vary widely, in accordance with the composition of the hydrothermal fluids and the associated country rock. Both sources are believed to contribute to the content of hydrothermal deposits.

Many metals are found in hydrothermal ores, such as copper, lead, zinc, gold, mercury, tin, tungsten, molybdenum, silver and uranium. Since sulfur is common in hydrothermal fluids derived from igneous and sedimentary sources, many of the minerals created are sulfides, such as galena (PbS) and sphalerite (ZnS). Where the ore-forming solutions originate from connate or meteoric waters, carbonates, sulfates (which can be oxidized to sulfur) and other anions are combined with metals.

4.3 Composition of Hydrothermal Solutions

The main component of hydrothermal solutions is water. Pure water, however, cannot dissolve metals, therefore, hydrothermal solutions are always brines, containing dissolved salts such as NaCl, KCl, $CaSO_4$ and $CaCl_2$. The range in salinity varies from that of seawater (around 3.5 wt. %) to about ten times the salinity of seawater. Such brines are capable of dissolving small amounts of metals such as Au, Ag, Cu, Pb and Zn. High temperatures increase the effectiveness of the brines to dissolve metals. The waters of hydrothermal solutions can be derived from several sources including magmatic fluids, seawater and meteoric (ground water) circulating through fractured rock, and connate brines (water trapped within sediments at deposition). Meteoric and seawater can form hydrothermal solutions if they are heated sufficiently and a convection system is generated. The source of this heat could be magmatic intrusions or local geothermal gradients. Regional or contact metamorphism can create hydrothermal solutions by dehydration of hydrous minerals during metamorphism or remobilization of connate water.

4.4 Conditions for Formation of Hydrothermal Deposits

Most hydrothermal solutions are believed to originate from deep in the upper mantle or crust in association with magmatic, metamorphic and sedimentary processes. Hydrothermal deposits are epigenetic, which means they were emplaced after the host rocks were formed. The required open spaces for ore localization must be available in the host rocks. As heated hydrothermal fluids carrying metals migrate upwards within the crust, temperature falls and the dissolved minerals precipitate out of solution. To be effective in generating sufficient mineralization to form ore bodies, the mineralizing process must continue over a long period of time and the solutions must move rapidly into porous layers or open spaces (e.g. fractures) which may lead to sudden cooling and rapid precipitation of minerals over a limited region.

The main reason why ore-bearing hydrothermal solutions deposit minerals is the loss of physical and/or chemical stability due to one or more of the following; boiling of hydrothermal solution, falling temperature, loss of pressure, reaction with wall rocks and degassing of the hydrothermal solution (separation into gas and water phases) which alters the metal carrying capacity of the fluid. Hydrothermal deposits are formed at a wide range of depths, from near surface to up to 10km in depth. The temperature of formation ranges from about 550°C to as low as 25° C.

The following conditions are conducive for the formation of hydrothermal mineral deposits: (1) presence of hot water to dissolve and transport minerals; (ii) presence of interconnected openings in the rock to allow the solutions to migrate; (iii) availability of sites or space for the deposits, and (iv) chemical reaction that will result in mineral precipitation.

4.5 Types of Hydrothermal Deposits

Hydrothermal deposits can best be classified according to their mode of occurrence which is largely determined by the structural and stratigraphic features that control their localization. Based on the mode of occurrence and environment of deposition, the following types of hydrothermal deposits are described and illustrated in Figure 4.1. Table 4.1 lists some characteristics of hydrothermal deposits.

1) Vein Deposits
2) Porphyry Deposits
3) Volcanogenic Massive Sulfide (VMS) Deposits
4) Sedimentary Exhalative (SEDEX) Deposits
5) Epithermal-Geothermal Deposits
6) Mississippi Valley Type (MVT) Deposits

Table 4.1: Summary Characteristics of Hydrothermal Deposit Types

Hydrothermal deposit Subcategory	Known abbreviation	Formation	Principal metals	Host rocks
Porphyry	—	Epigenetic	Cu, Mo, Au	The ore is spatially associated with one or more high-level intrusions of felsic to intermediate composition such as granite, granodiorite or diorite
Vein Deposits	—	Epigenetic	Cu, Mo, Ag, Au, W, Sn	Associated with wide variety of rocks including igneous, sedimentary and metamorphic
Volcanogenic Massive Sulfides	VMS	Epigenetic/Syngenetic	Cu, Zn, Pb	The host rocks are mainly volcanic, with the felsic volcanic rocks pointing to a convergent setting (island arc or orogenic belt).[5]
Sedimentary Exhalative	SEDEX	Syngenetic	Zn, Pb	These deposits are commonly stratiform and are typically hosted in marine hale.
Epithermal Geothermal	—	Epigenetic	Au, Ag	The host rocks can be sheared muscovite granite, small plutons.
Mississippi Valley-type	MVT	Epigenetic	Pb, Zn	MVT deposits are hosted in carbonate rocks,

4.5.1 Vein Deposits

Veins (vein deposits) are the most widespread or classical form of hydrothermal deposits. They occur as tabular, elongated bodies that occupy openings in rocks within a fairly well-defined zone of mineralization. Often, they are usually vertical or inclined and discordant, but typically narrow when compared to its length and depth. Veins can be small or extensive, reaching a length of several hundred meters and a width of a few centimeters to tens of meters. Most vein deposits occur in fault or fissure openings or in shear zones within country rock. Their mode of occurrence is epigenetic, and therefore, may cut across a host of country rocks. A vein deposit that is rich in ore minerals is sometimes referred to as a (metalliferous) *lode* deposit. Veins may also occur singly, or as a system (vein system) composed of several veins that are conjugate and parallel (Fig. 4.2a). Small veins or veinlets may occur as *stockwork*, which are a small group of closely spaced and structurally controlled or randomly oriented veins and veinlets (also known as stringers) that form an interlacing network. Stockworks are common in many hydrothermal ores but are most prominent in greisens of tin deposits and vein systems associated with porphyry copper and molybdenum deposits. Temperatures of formation of vein deposits range from less than 50° to 500°, but most vein deposits are mesothermal, formed at temperatures between 200° to 300°C.

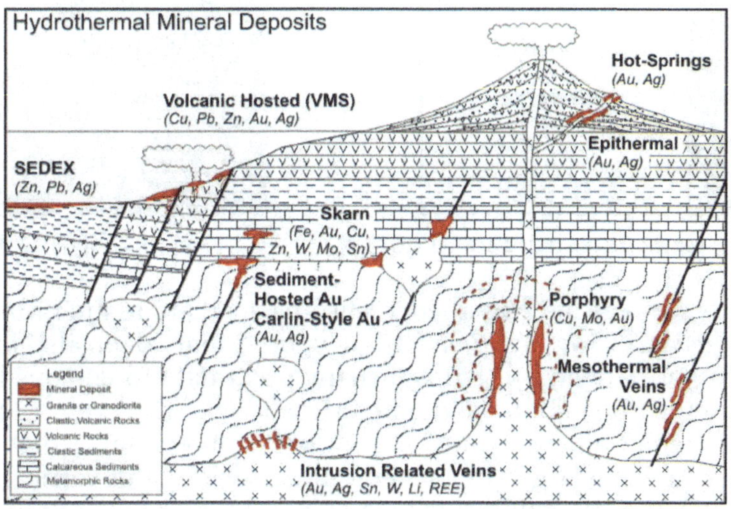

Figure 4.1: Schematic diagram of hydrothermal deposits

Vein deposits usually contain ore minerals with quartz and sometimes carbonate as gangue. They are sources of a variety of valuable metallic minerals, such as native gold or silver, cassiterite (Sn), wolframite (W), pitchblende (U) or metal sulfides of copper, molybdenum, lead, zinc and mercury, that are usually deposited along with gangue minerals (mainly quartz and calcite). Hydrothermal veins are commonly composed of quartz veins (containing metallic minerals) that occur as interlocking crystals in a variety of sizes or as finely laminated bands parallel to the walls of the vein typical of cavity filling texture. Vein deposits produce most of the world's copper and molybdenum, and also important for gold, tin, silver, mercury, and uranium (Fig. 4.2b). Veins are usually accompanied by wall rock hydrothermal alteration whose intensity decreases outwards from the center of the vein. Typical hydrothermal alteration around veins involves formation of clay minerals (argillic), sericite (phyllic) or sericite and chlorite (greisenization) that is common around tin lodes.

Some well-known examples of vein deposits are the Sn-Cu-Pb-Zn deposits in Cornwall, England, the gold-quartz veins of Kalgoorlie, Western Australia, and Kirkland Lake, Ontario, Canada, and the silver-nickel-uranium veins of the Erzgebirge, Germany. In Nigeria, hydrothermal vein deposits of gold occur at various locations within the Precambrian schist belts: including Iperindo (now Segilola), near Ilesha in southwest Nigeria, and in many other locations in north central (Birnin Gwari) and northwestern Nigeria. In the Benue trough, lead-zinc, barite and copper vein deposits occur within Cretaceous sediments near Abakaliki, Zurak, Azara, Wase and other localities in Benue, Nasarawa and Ebonyi states. Tin-bearing lodes have been mined in the Jos Plateau, and one of the prominent locations is the Ririwai lode in Kano State. These veins are associated with the Jurassic Young Granite complex.

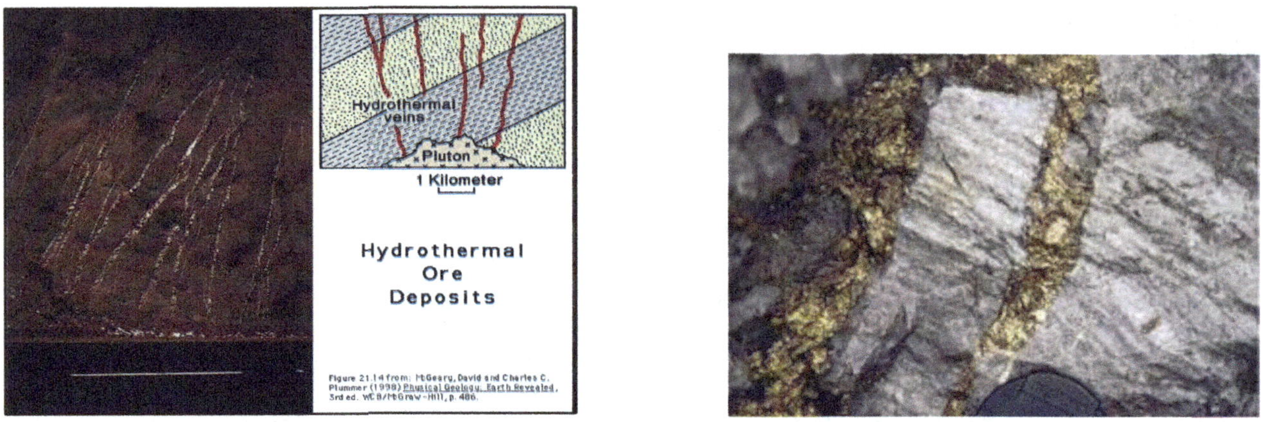

Figure 4.2: (a)Quartz veins and stringers in host rocks (b) Gold veinlet

Hydrothermal veins can be classified according to their temperature of formation into: *hypothermal, mesothermal* and *epithermal.* Hypothermal deposits are formed at high temperatures (300° to 500°C) and pressures, usually proximate to igneous bodies. Examples are cassiterite (Sn), wolframite (W) and molybdenite (Mo)and gold (Au) veins. Mesothermal veins are formed at moderate temperatures (200° to 300°C) and intermediate depths. They are characterized by their relatively large size (several meters wide) and vertical extent. Ore minerals include native gold (Au) and the sulfides galena (PbS), sphalerite (ZnS), chalcopyrite ($CuFeS_2$), pyrite (FeS_2), and bornite (Cu_5FeS_4). Epithermal veins are usually short and narrow, and localized within irregular branching fissures, stockworks or form at shallow depths under relatively low temperatures (50° to 200°C) and pressures. Colloform and replacement textures are sometimes observed, but the majority of epithermal vein deposits are characterized by open-space filling textures (crustification, comb structures, symmetrical banding) – all indicators of low temperatures of formation. Ore minerals are native gold (Au); and the sulfides; realgar (AsS), cinnabar (HgS), acanthite (Ag_2S), pyrite (FeS_2), orpiment (As_2S_3), stibnite (Sb_2S_3),), and pyrargyrite (Ag_3SbS_3) and barite ($BaSO_4$).

The origin of vein deposits is well known. They are believed to have formed by hydrothermal solutions emanating from surrounding igneous intrusions or other sources and injecting metals and silica into open fractures. The ore-forming fluids cool rapidly and precipitate mainly quartz, but also a variety of sulfide minerals, and sometimes, gold and silver within the veins of quartz. Rich deposits of copper, zinc, lead, gold, silver, tin, tungsten, mercury, and molybdenum are commonly found in vein deposits.

4.5.2 Porphyry Deposits

The term "porphyry" deposits refers to a group of large and low-grade, intrusion-related deposits comprising veins, stockworks and dissemination of sulfide minerals in granitic intrusive rocks that are porphyritic in texture. While the term porphyry-type is more widely used to describe copper deposits, porphyry molybdenum, gold, tin and tungsten deposits have also been reported. These deposits are mostly Tertiary in age and found along the western continental margins of North and South America, particularly in the Cordilleran region of British Columbia, Canada, and in Arizona, Montana, and Utah in the United States. The intrusive rocks associated with porphyry-type deposits range in composition from quartz diorite to monzonite, and occur as plug-like stocks, with swarms of dykes and intrusive breccias. When formed in differentiated intrusive sequences, mineralization tends to be localized in the finest-grained and most felsic end members of the suite. Porphyritic rocks and aplites usually form part of the intrusive suite (Fig. 4.3).

The porphyry mineralization is obviously epigenetic, and occurs in narrow, closely-spaced veins and veinlets within hydrothermally altered rock. This produces stockwork ore, as well as disseminated ore, which is actually mineralization along numerous, closely spaced micro-fractures, although there is often evidence of dissemination in altered rock outside the vein system. The major ores of porphyry copper deposits are copper with minor gold. Chalcopyrite is the dominant ore mineral, often with minor bornite, chalcocite and covellite. Secondary minerals that develop in the zone of supergene enrichment are malachite and azurite. The main gangue sulfide mineral is pyrite, and magnetite is the main oxide mineral. Porphyry molybdenum deposits are not common, but they are the main source of molybdenum worldwide. The Climax Mine in Boulder Colorado, USA supplied over 75% of global molybdenum production. Many molybdenum deposits contain some tin and tungsten minerals which are recovered from the ores for additional value. Porphyry tin deposits have been described from several localities in Bolivia, South America where high-level granitic stocks contain stockwork, disseminated and breccia-filling cassiterite.

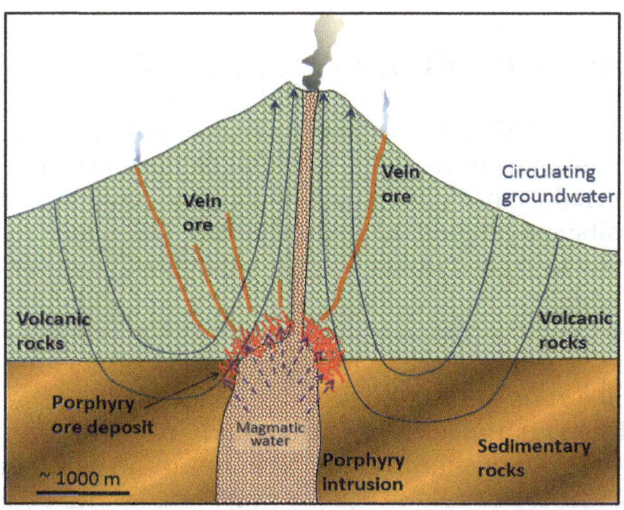

Figure 4.3: Schematic diagram of porphyry-type ore deposit

Hydrothermal alteration is commonly associated with porphyry type deposits; it is very distinctive, and changes as you approach mineralization. In some large deposits, the wall rock alteration is widespread and extensive, sometimes up to several kilometers wide, and is typically zoned on a deposit scale, as well as around individual veins and fractures (Figure 36). The most common types of hydrothermal alteration are propylitic (chlorite), argillic (clays), phyllic (sericite) and potassic (k-feldspar) (Fig. 4.4). Where sulfide mineralization occurs on the surface, weathering and oxidation often produces rusty-stained bleached zones rich in Fe hydroxides called gossans, from which the metals have been leached.

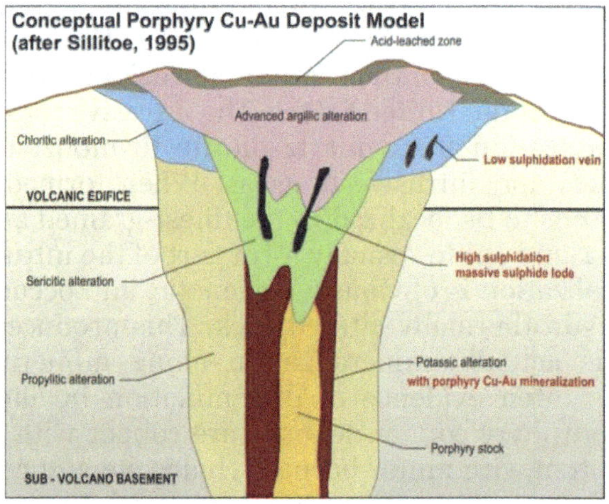

Figure 4.4: Schematic diagram of hydrothermal alteration in porphyry-type deposit

The origin of porphyry-type deposits is a bit complex. Fluid inclusion studies show that hydrothermal fluids ranging from magmatic to meteoric sources have been involved in the formation of porphyry-type deposits. Various factors, such as magma type, volatile content, the number, size, timing and depth of emplacement of mineralizing porphyry plutons, variations in country rock composition and fracturing, all combine to ensure a wide variety of detail in this group of hydrothermal deposits. A conceptual model of porphyry deposit is illustrated in Fig. 4.4.

4.5.3 Volcanogenic Massive Sulfide (VMS) Deposits

Volcanogenic massive sulfide (VMS) deposits are metallic sulfide-rich (>60% sulfides) deposits hosted or associated with submarine mafic to felsic volcanic rocks. They are deposits created by volcanic-related hydrothermal events in submarine volcanic environments. The main reason why these intra-volcanic deposits are considered hydrothermal is because mineralization is epigenetic and deposited in volcanic rocks by hot fluids, mostly seawater that was mobilized by post-volcanic activity.

VMS deposits have the following unique characteristics: (1) they are closely associated with volcanic rocks; (2) the ores are "massive" (with over 60% sulfides) and stratiform (or nearly so); and (3) they have a pipe-like zone of stockwork veins localized along the conduit zone beneath the massive ores (Fig. 4.5). Deposits range in age from Early Archean (3.55 by) to the present, and significant occurrences of VMS mineralization are found in greenstone belts of almost all Precambrian shield areas. Mineralization is usually polymetallic and composed of the base metals; copper, lead and zinc (Cu-Pb-Zn) with minor gold and silver. These deposits are believed to have been created by volcanic-related, hydrothermal events in submarine environments on the seafloor. The three recognized types of VMS deposits named after the type areas are called Kuroko, Cyprus and Besshi types. Characteristic host rocks are of three types: (1) *felsic lavas* (Kuroko type), (2) *mafic lavas* (Cyprus type) and, (3) *mixed volcanic/pelitic* sediments (Besshi type). VMS mineralization occurs mainly as lens-shaped bodies of massive ore underlain by a stringer ore zone consisting of stockwork-like veins, as well as pod-like replacements of the host rock. The stringer zone is roughly perpendicular to the massive ore and thus transects the volcanic host rocks. It narrows with depth and grades upwards into the massive ore.

The VMS deposits are important sources of the world's copper, zinc, lead, gold, and silver (Cu, Zn, Pb, Au, and Ag) ores. The different ore associations based on type of host rock in VMS deposits are: 1) felsic volcanic-hosted Kuroko type- Cu-Pb-Zn-Ag-Au; 2) mafic-volcanic hosted Cyprus type - Cu (-Zn, -Au); and 3) mixed volcanic/sedimentary Besshi type - Cu-Zn (-Au). The ore minerals are abundant iron sulfides (pyrite or pyrrhotite) with variable but subordinate amounts of chalcopyrite and sphalerite; bornite, tetrahedrite, and galena. Gangue minerals typically consist of quartz, barite, anhydrite, iron oxides, chlorite, sericite, talc, and their metamorphosed equivalents. Vertical zoning of the ore minerals within the volcanic pile from Fe-Cu at the base to Cu-Fe in the main ore zone and Zn-Pb-Fe at the top (Fig, 3.28). The size of the deposit ranges from small pods of less than a ton to supergiant accumulations such as the Rio Tinto deposit in Spain with over 1.5 billion metric tons (Bt) of ore.

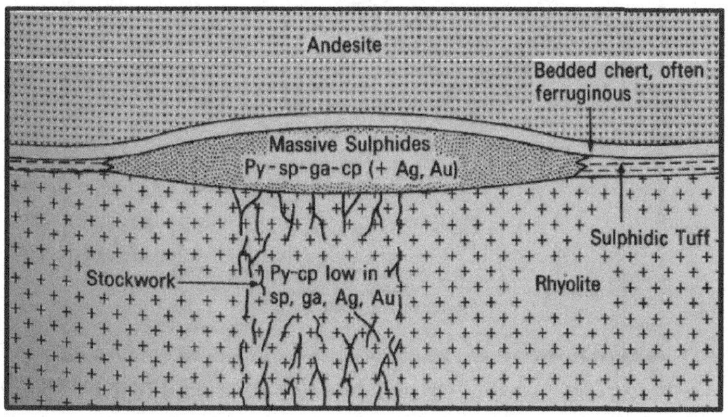

Figure 4.5: Volcanogenic massive sulfides showing stockwork feeder zone

Zones of hydrothermally altered rock are associated with all types of VMS deposits. These alteration zones characterized by silicate minerals (chlorite, Illite, sericite) are very distinctive, and cut across the volcanic stratigraphy. As shown in Fig. 4.6), the conduit pipe zone with stockwork mineralization beneath the massive sulfide orebody is usually intensely altered. The alteration was produced by the interaction of a hot (metal-rich) fluid with the volcanic or volcaniclastic wall rocks. The classic VMS deposit is called the "Kuroko-type", named after the type locality in northeast Japan. In the deposit, the felsic lavas are associated with explosive eruptions which laid down pyroclastic rocks that were interspersed with layers of marine sediments. At the vent locations where the lavas erupted, there is typically a plug or dome shaped intrusion of rhyolite or dacite composition. Mineralization occurs in the form of massive sulfides or dense concentrations of disseminated sulfide minerals of various types. Two types of ore are recognized at Kuroko: "yellow ore" which occurs as abundant massive pyrite hosted in brecciated tuffaceous rocks and. "black ore" which is characterized by concentrations of finer-grained sphalerite and galena (Fig. 4.7).

In terms of origin, the close association of VMS deposits with volcanic rocks suggests that ore deposits were formed at or near the sea floor during submarine volcanism. The circulation and mixing of two types of solutions, seawater and "juvenile" water (from the subjacent magma chamber) is thought to play a major role in ore formation and distribution of ore metals. A sub-seafloor hydrothermal system driven by heat from the emplacement of subjacent mafic intrusions mobilized heated connate seawater that rose and leached metals from the volcanic pile and precipitated the ores as the rising solutions interacted with cold seawater. The presence of pervasive alteration zones characterized by secondary quartz and phyllosilicate minerals are indicative of hydrothermal circulation through footwall volcanic rocks which strongly support the epigenetic origin of VMS deposits.

Fluid inclusion studies indicate high salinity brines and temperatures ranging from 300 - 400 degrees centigrade. Commonly, there is an apparent zonation of ore metals from a central zone of copper-rich mineralization (yellow ore) to an outside zone of lead and zinc-rich mineralization (black ore) (Fig. 3.29). Barite ore forms the most distal portions of the zonation. The metals in the deposits were derived from leaching of the volcanic rocks and related magmatic sources.

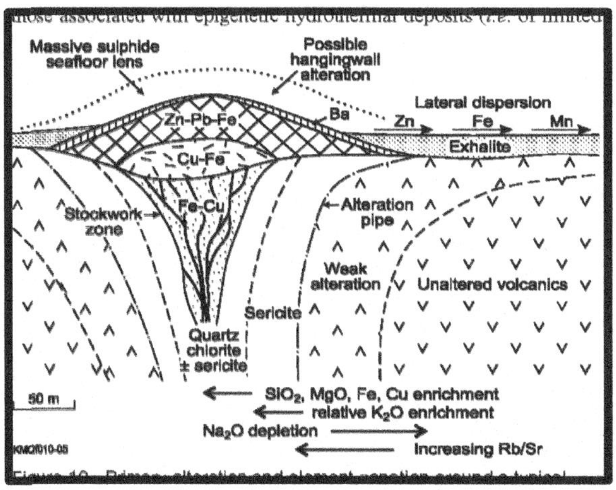

Figure 4.6: Metal and alteration zoning in idealized VMS deposit.

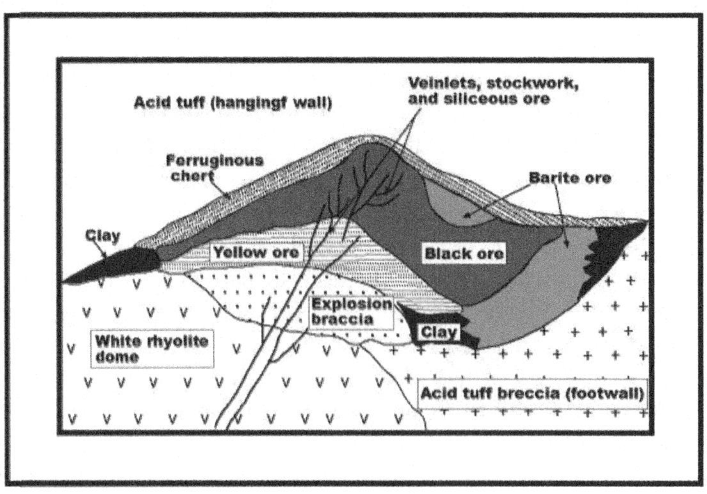

Fig. 4.7: Geology and ore distribution at Kuroko-type VMS deposit

4.5.4 Sedimentary Exhalative (SEDEX) Deposits

Sedimentary exhalative deposits (SedEx) are mineral deposits of Pb-Zn-Ag formed by precipitation of metals into sea water, through the release of ore-bearing hydrothermal fluids, resulting in the formation of sediment hosted (stratabound) mineral deposits. The term "exhalative" refers to the process of "venting hydrothermal solutions into submarine environments" and concentrating the ores in sediments. SedEx deposits are the most important source of lead, zinc and barite, and account for more than 50 percent of the world's zinc (Zn) and lead (Pb) reserves) and furnish more than 25% of the world's production of these two metals. Examples of SedEx deposits are Broken Hill and Mt. Isa, Australia; Howards Pass, Yukon and Sullivan mine, British Columbia, Canada.

SedEx Pb-Zn-Ag deposits are bedded or laminated, tabular, sulfide–rich bodies in carbonaceous shales (black shales and mudstones) or other fine-grained clastic rocks deposited in deep, usually intracratonic rift basins. Less commonly the deposits are hosted in limestone units forming shelf-like layers near the edges of the basins. Limestone hosted deposits tend to occur more as veins and be associated with breccias. Banded or layered deposits, semi-concordant to the stratigraphic layering, are generally rich in zinc. The sphalerite occurs as dense accumulations of fine-grained sulfides in parallel layers or bands. This banding is generally parallel to layering in the overall stratigraphy.

Sediment hosted exhalative deposits demonstrate the following unique characteristics compared to other similar deposits: 1) they show no spatial, temporal, or genetic association with volcanic rocks: 2) they typically form on the ocean floor, usually around the edges of large, deep sedimentary basins: 3) are characterized by having very high zinc, lead, silver and barium values: 4) zinc and lead occur mainly as sphalerite and galena, and silver occurs primarily as argentiferous galena. SedEx deposits are best known as major sources of lead-zinc ores worldwide, and produce the vast majority of the largest and most important deposits Moreover, other forms of SedEx mineralization are known; such as, the vast majority of the world's barite deposits are considered to have been formed by SedEx mineralization processes.

The mode of origin of SedEx deposits is attributed to precipitation of metals through submarine venting of hydrothermal fluids in sedimentary basins. Based on fluid inclusion studies, the temperature of ore formation ranges from 100 – 300°.C, and the ore-forming solutions are heated saline brines. Mineralization is essentially syn-sedimentary; forming at the same time as the host rocks or just below the seafloor within fault-controlled basins or troughs that are most often related to major intracratonic or continental margin rift zones (Figure 42). High angle fault structures, near the edges of the basin, appear to have channeled metal rich, hot water brines. As the fluids migrated away from the feeder zone, chemical and physical changes of the brines resulted in precipitation of the metals.

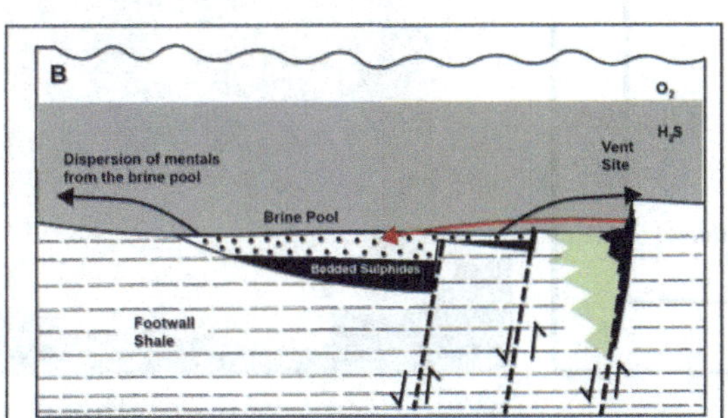

Figure 4.8: Schematic diagram of SedEx deposits

(b) Drill cores of laminated SedEx deposits

4.5.5 Epithermal- Geothermal Deposits

Geothermal-Epithermal (geo-epithermal) deposits are a small group of hydrothermal deposits of Au-Ag closely associated with volcanic geothermal fields and formed at shallow depths and at low pressure and temperature. They are attributed to the action of heated groundwater (hot springs) associated with geothermal sources. They are low sulfide deposits composed mostly of gold, silver, antimony and mercury. Temperature of formation is in the range of <150 to 300°C, and are often described as epithermal deposits. Fluid inclusion studies, the ore mineralogy and related textures indicate that these deposits were formed at shallow depths (<2km).

Host rocks are mostly volcanic rocks with subaerial pyroclastic rocks and sub-volcanic intrusions. Hot springs and fumarole deposits (e.g. sulfur) may be present in deposits that are not deeply eroded. Gold and silver are the main economic minerals in geothermal -epithermal deposits with lesser amounts of mercury, arsenic and antimony (Fig. 4.9b). Ore minerals are native gold, argentite (Ag), tennantite (As), stibnite (Sb), and cinnabar (Hg). Gangue minerals are mainly quartz and calcite with lesser fluorite, barite, rhodochrosite, dolomite and chalcedony. Geothermal gold deposits are found in New Zealand, Tasmania and near Sulfur in Nevada and other localities in Colorado, USA. New Zealand has several commercial epithermal gold deposits, but is also a country where the creation of geothermal gold deposits can be observed today in the zone of hot springs and geysers in parts of North Island, New Zealand. Epithermal-Geothermal deposits are believed to originate from low temperature geothermal hydrothermal systems related to volcanic activity.

These systems, while active, discharge to the surface as hot springs or fumaroles. This is evidenced by the fact that many geothermal deposits appear to be spatially associated with hot springs and geysers, and these hydrothermal systems may be considered the surface expression of heated groundwater systems. Apparently, hydrothermal solutions can be formed by heated groundwater due to geothermal activity near shallow subvolcanic environments (Fig. 4.9a). Fluid inclusion studies indicate that these geothermal deposits form from dilute (< 5 wt % NaCl) waters that undergo boiling or effervescent degassing, fluid mixing and oxidation at temperatures in the range of 200 - 300°C. Boiling and mixing of fluids appears to be the most important way in which these hydrothermal solutions cool down. And deposit economic minerals.

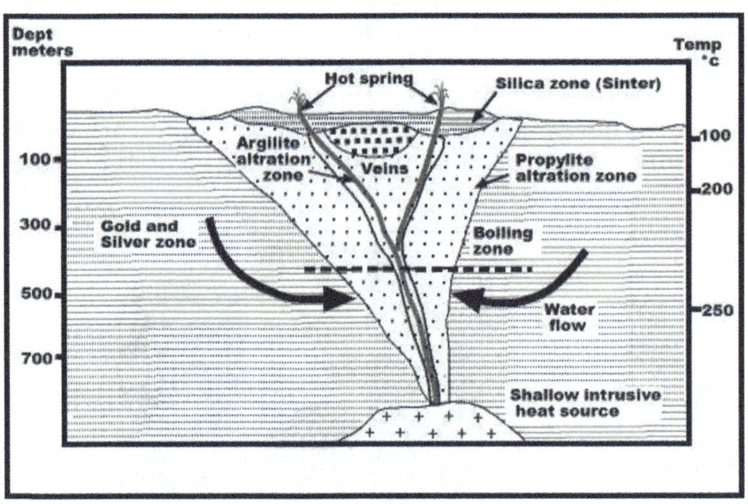

Figure 4.9: Diagram of geo-epithermal deposits

(b) Gold-silver geo-epithermal ore

4.5.6 Mississippi Valley Type (MVT) Deposits

Mississippi Valley-Type (MVT) deposits are a varied family of epigenetic lead-zinc deposits that occur predominantly in carbonate host rocks, particularly dolomites and limestones of mostly Paleozoic (Cambrian) to Mesozoic (Cretaceous) age. They derive their name from the Mississippi River in south-eastern United States where classic deposits occur. Mineralization is epigenetic but may in places be locally stratiform, and enclosed or parallel with surrounding sedimentary host rock (Fig. 4.10a). The most important ore controls are open spaces created by faults and fractures, dissolution collapse breccias, and lithological transitions. The deposits include tectonic or karst cavity breccias that were cemented by ore and gangue minerals; replacement bodies and cross-cutting veins. Texturally, the ores may be massive, banded or brecciated (Fig. 4.10b). The principal ore minerals are galena and sphalerite with minor silver, Gangue minerals may include carbonates with minor barite and some fluorite. Silicification of the carbonate host rocks may accompany mineralization.

Some of the unique characteristics of MVT deposits are the localization of ores in open spaces and permeable zones within platform carbonate sequences along flanks or edges of sedimentary basins where they appear parallel to the reefs. Sometimes the carbonate host rocks constitute a part of a reef, and the ores are localized relative to a particular reef facies. In other cases, the ores may occur in association with solution collapse breccias. Many of the larger orebodies occur as stratiform masses (i.e., parallel, or nearly so, with the bedding), and some as beds that extend from a few centimeters to several tens of meters with ore banding on a very fine scale. Elsewhere the deposits occur as vein infillings or replacements along fissures (commonly faults or joints) that may cross the bedding. These epigenetic features have created controversy about the origin of these deposits that are still considered by some authors as syngenetic-diagenetic. MVT deposits are found throughout the world but are most abundant in Europe and North America.

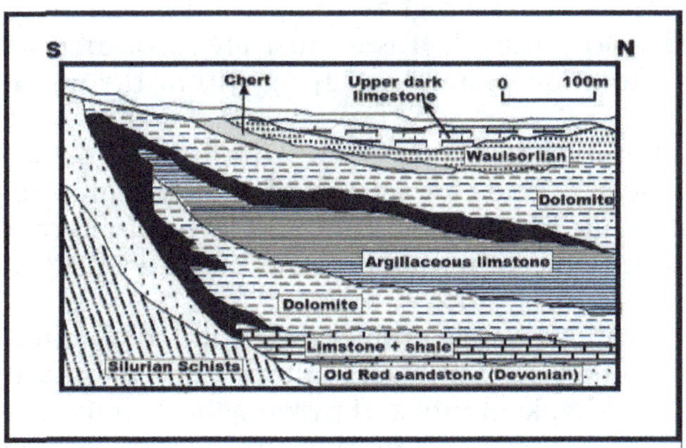

Figure 4.10: MVT Pb-Zn Deposits in Carbonates (b) MVT Pb-Zn Ore with colloform bands

4.6 Wall Rock Alteration in Hydrothermal Deposits

Wall-rock alteration is one of the unique features of hydrothermal deposits. All types of hydrothermal deposits are associated with one form or the other of wall rock alteration which is attributed to interaction of host rocks with mineralizing hydrothermal solutions. Notable deposits with widespread alteration zones are the porphyry type, volcanogenic massive sulfide and geo-epithermal deposits. Wall-rock alteration occurs over a wide range of temperatures, and is often characterized by changes in the mineralogy of the country rock as a result of a chemical reaction with hydrothermal solutions.

The changes in mineralogy are most marked adjacent to the vein or center of deposit, and become less distinct further away. It may take place in advance of emplacement of the ore minerals, or during the final waning stages of hydrothermal activity. An alteration zone defines rocks that have been altered to a specific group or assemblage of secondary or alteration minerals, usually around the perimeter of a mineral deposit. However, in large hydrothermal deposits such as porphyry type, the alteration zones may be pervasive or cover a large area around the mineralized zone. Chemical changes in the various types of alteration are summarized in Table 4.2. Hydrothermal alteration has been recognized as a valuable tool in mineral exploration, because the alteration zones around many deposits are wider and easier to locate than the ore bodies themselves.

4.6.1 Types of Hydrothermal Alteration

Types of wall rock alteration are identified by the type of mineral or mineral assemblage present in the alteration zone. Hydrothermal alteration can be described as "major" or "minor" depending on the areal extent of the alteration. The most common alteration types are: (1) Potassic; (2) Argillic; (2) Phyllic; (3) Propylitic and (4) Greisinization. Other minor forms of hydrothermal alteration include: chloritization, silicification, feldspathization, carbonatization, and tourmalinization. Zoning of wall rock alteration is common around vein deposits and porphyry copper deposits (Fig. 4.11),

Potassic Alteration: is characterized by the presence of new K-Feldspar and/or biotite, sometimes with minor sericite and quartz. It is sometimes distinguished by the pink to red coloration of the altered rock. Potassic alteration represents the highest temperature encountered in hydrothermal processes, usually in the range of 500 – 600°C. It is exclusively associated with porphyry-type deposits, forming the core of the alteration system and lying within the granitic intrusion.

Phyllic Alteration: is the most common alteration in a variety of hydrothermal deposits and forms over a wide temperature range of 200 – 500°C, and characterized by the assemblage of sericite (a fine-grained muscovite) and quartz. It is one of the most common form of alteration around porphyry type base metal deposits, but also with mesothermal precious metal deposits and VMS ores in felsic rocks (Kuroko type).

Argillic Alteration: is characterized by the presence of clay minerals (hydrated aluminum silicates) and quartz. There are two types based on mineral assemblage: intermediate argillic and advanced argillic. In *advanced argillic alteration*, dickite, kaolinite and pyrophyllite (all hydrated aluminum silicates) and quartz are the alteration assemblage. Alteration involves the extreme leaching of cations, especially the difficult to leach alkalis and calcium, and the concentration of H+. This type of alteration is characteristic of many epithermal precious metal deposits. *Intermediate argillic alteration* is characterized by kaolinite and montmorillonite and is somewhat similar to advanced argillic alteration, but with a lesser degree of leaching of cations. Unlike advanced argillic alteration which is associated with vein type deposits, intermediate argillic alteration is more closely associated with disseminated deposits such as porphyry coppers.

Propylitic Alteration: is probably the most widespread type of alteration. It is characterized by the assemblage chlorite + epidote + calcite. Albite, as well as other carbonates may be present. Due to the presence of the green minerals, chlorite and epidote, this zone is usually easily recognizable by its green color. Unlike the previous types above which are characterized by leaching of cations, this zone seems to represent the addition of cations of calcium and magnesium. Propylitic alteration is common in volcanic host rocks of vein and porphyry type deposits, as well as beneath VMS deposits.

Greisenization: is the hydrothermal alteration of granitic rocks into an assemblage of quartz + mica + chlorite + tourmaline + topaz + (fluorite). It is believed that this type of alteration is aided by the action of water vapor containing fluorine. The associated metallic minerals are cassiterite + various tungsten-bearing minerals. This is a common form of alteration in association with tin deposits in granites.

Tourmalinization: is the introduction of tourmaline crystals and quartz- sericite – tourmaline veins into altered granitic rocks and pegmatites at medium to high temperature. This type of alteration is most commonly associated with mineralization in granitic rocks.

Silicification: is the addition of new quartz or amorphous silica minerals (such as chalcedony) to a rock during alteration. This type of alteration is common in hydrothermal vein deposits.

Hematitization: this is alteration that is characterized by the formation of minerals with a high $Fe3^{+}/Fe2^{+}$ ratio, particularly the introduction of hematite as a result of oxidizing hydrothermal fluids. In porphyry copper deposits hematite is at times associated with potassic alteration.

Chemical changes associated with hydrothermal alteration are summarized in Table4.2. Potassic alteration and Na metasomatism which represent relatively high temperature forms of alteration show increased concentrations of potassium and sodium in K-feldspathization and albitization respectively. There is a corresponding loss of other elements, Ca, Mg and Fe. Phyllic alteration is characterized by the greater abundance of K-mica, sericite and muscovite and quartz. The chemical changes involve the addition of k and the loss of Ca, Na and other elements. Argillic alteration is dominated by clay minerals which results in the loss of both K and Na, and other elements and increase in H and C. Propylitic alteration and chloritization occur at relatively lower temperatures, and involve the addition of Fe, Mg and Ca.

Table 4.2: Alteration types and associated chemical changes and alteration minerals.

	Type of Alteration	Mass Changes	Examples of Mineral Assemblages
	Potassic (K-Feldspar) Alteration	+K, Na	K-feldspar + biotite + quartz
	Albitization	+Na, -K	Albite + hornblende ± biotite-quartz
	Phyllic (sericite) Alteration	+K or -K, -Na, -Ca, +H	White mica + quartz + pyrite
	Argillic clay Alteration	-Na, K, -Ca	kaolinite, illite or montmorillonite
	Propylitic	+H, +C	Epidote + chlorite + albite ± carbonate
	Chloritization	+Fe, +Mg, +H -Na, +Ca, -K	Chlorite + pyrite + white mica ± quartz
	Silicification	+Si	Any minerals + quartz
	Carbonatization	+C, +Ca	Carbonates, quartz-white mica-chlorite Talc + chlorite + car

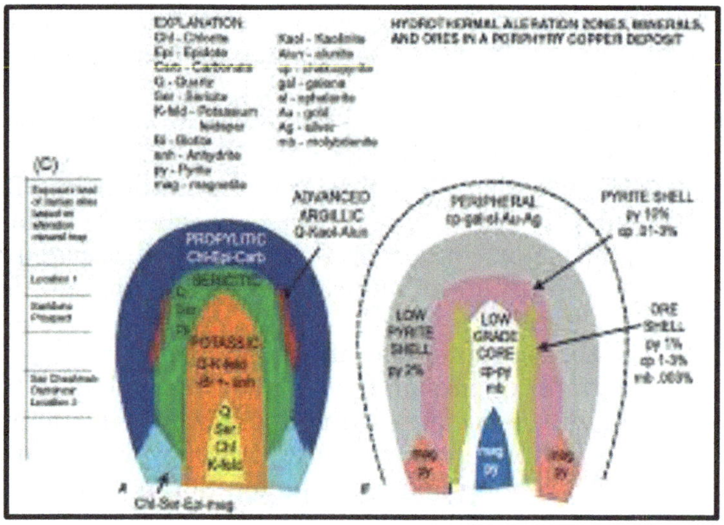

Fig. 4.11: Zoning of wall rock alteration around porphyry deposits

Revision Questions

1. Describe the sources, composition and transportation of hydrothermal solutions.
2. What are hydrothermal deposits? Attempt a classification based on the temperature of deposition.
3. Differentiate between the following pairs of terms: (a)veins and stockworks; (b) epigenetic and syngenetic; (c) potassic and argillic alteration; (d) mesothermal and epithermal deposits.
4. Describe the main features and differences between volcanogenic massive sulfide (VMS) and sedimentary exhalative (SEDEX) deposits,
5. Write short notes on two of the following: (a) greisenization, (b) tourmalinization and (c) Mississippi Valley type deposits.

Chapter 5: Sedimentary Mineral Deposits

5.1 Introduction

Sedimentary mineral deposits are mineral concentrations produced by sedimentation or sedimentary processes. Sedimentary environments are quite varied and complex, ranging from fluvial, lacustrine, coastal, and shallow to deep marine. The processes leading to the concentration of mineral deposits may stretch across several contiguous environments. Metals are usually released from the land masses by physical or chemical weathering and transported across the land by erosion and surface water to the sedimentary depocenters. The stability of dissolved metals is surface media is determined by a combination of physical, chemical and biological factors, and their complex interactions. The materials that are found in sedimentary mineral deposits have been derived mostly from weathering of existing rocks, with minor contributions from extraneous discharge of metal-bearing fluids into sea water, and subsequent diagenetic remobilization.

The essential conditions for the formation of sedimentary mineral deposits include the (1) adequate source of ore materials: (2) gathering of the materials by dissolution or other processes: (3) transportation of the materials to the site of accumulation, (4) deposition of materials in the sedimentary basins; and (5) compaction and diagenesis. Sedimentary deposits are major sources of some important metals including, iron, manganese, copper, lead, zinc, uranium, gold, diamond, coal and several industrial mineral deposits, such as carbonates, phosphates, potash, celestite, barite and bentonite.

5.2 Types of Sedimentary Mineral Deposits

Sedimentary mineral deposits can be classified into five main groups based on the processes of ore formation and environment of deposition.

1) Placer deposits
2) Evaporite deposits
3) Chemical precipitate deposits
4) Sediment-hosted (Stratiform)) deposits
5) Biogenic deposits

5.2.1 Placer Deposits

The term "placer" is used to describe deposits formed by mechanical accumulation of valuable detrital and resistant minerals released by weathering from appropriate source rocks and concentrated on slopes, riverbeds or beaches (Fig. 5.1). Minerals that are commonly concentrated in modern placer deposits are heavy and stable minerals that may remain in place or slowly washed down slopes into streams where they are further segregated. Most placer deposits are found in alluvial or beach gravels and sands that are geologically young in age, although old placers, known as "paleo-placers" have been reported in indurated sediments. Notable examples of placer minerals are diamond, gemstones, some native metals (gold and platinum-group elements), and oxides (uraninite, chromite, magnetite, ilmenite, rutile, cassiterite, wolframite, columbite-tantalite), There are certain preconditions for the concentration of placer minerals, including: (1) mechanical and chemical durability (2) resistance during weathering, erosion and transport, and (3) higher density compared to other minerals. Placers are important sources of diamond, gold, platinum, tin, titanium (rutile, ilmenite), zircon, rare earth elements (monazite) and gemstones.

Based on the geological environment of accumulation as shown in Figure 5.1, placers may be classified into various types namely: *eluvial* placer (material in weathering profiles), *colluvial* placer (transported along slope by gravity), *alluvial* placer (transported by a river*), beach* placer (deposited along the edge of water bodies) and *paleo*-placer (ancient placers that are consolidated or metamorphosed). In arid regions where there is no flowing water, *aeolian* placers are created by the winnowing action of wind. As shown in Figure 5.2, beach placers are usually finer grained and better sorted than alluvial placers which are formed by running water.

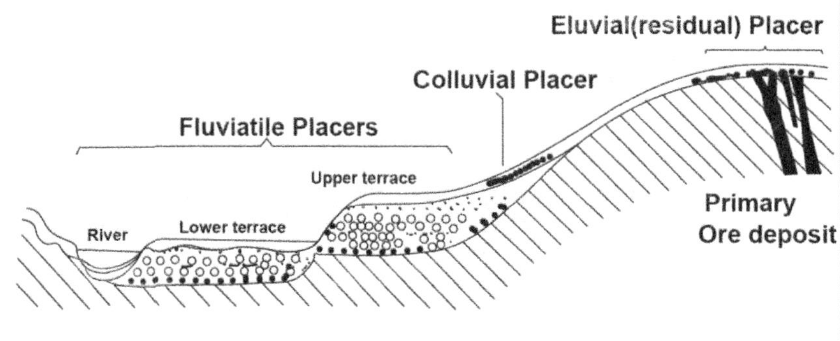

Fig. 5.1: Types of placer deposits

Placer deposits are important sources of gold, cassiterite (Sn), diamonds, zircon, ilmenite and rutile (sources of Ti). Examples are numerous, including the tin (cassiterite) deposits in Nigeria's Jos Plateau, Malaysia, Thailand, and Indonesia, and the diamond placers in Congo (Kinshasa), Angola, and Orange Free State, South Africa. Several gold placers also occur in West Africa including Nigeria, Ghana, Sierra Leone and Burkina Faso, and ilmenite is found in beach placers along the west and east coasts of South Africa and Mozambique, Paleo-placer deposits of gold and uraninite occur in pebble conglomerates of the Witwatersrand Supergroup, South Africa; Elliot Lake (Blind River), Ontario, Canada (major uranium deposit); and in Jacobina, Brazil, (gold).

Fig. 5.2: (a) Alluvial placer deposit, South Africa (b) Beach placer mineral sand, India

5.2.2 Evaporite Deposits

Evaporite deposits are those deposits formed by evaporation of lakes or sea water which results in the concentration of dissolved substances that precipitate from the water. The most common evaporite deposits are salts (most commonly sylvite and halite), gypsum, potash and nitrates. Evaporite deposits usually occur as thick beds or thin layers of salts that accumulate in thick sequences as sedimentary rocks. There are two types of evaporite deposits; 1) *non-marine* (lakes) and 2) *marine* (sea or ocean). Non-marine evaporites that precipitate from lake water of suitable composition include sodium carbonate, sodium sulfate and borates. Huge deposits of sodium carbonate occur in the Green River basin of Wyoming, USA. Borax and other boron-containing minerals are mined from evaporite lake deposits in Death Valley and Searled and Borax Lakes, all in California; and in Argentina, Bolivia, Turkey, and China. Lake Chad in northeastern Nigeria is associated with evaporites including trona, a hydrated carbonate and bicarbonate of sodium, and other salts formed by the evaporation of the waters of the historic Lake Chad.

Much more common and important than non-marine evaporites are the marine evaporites formed by evaporation of seawater. The most important salts that precipitate from seawater are: Gypsum ($CaSO_4.2H_2O$). Halite (NaCl) and Carnallite ($KCl.MgCl_2.6H_2O$). Low-grade metamorphism of marine evaporite deposits causes another important mineral, sylvite (KCl), to form from carnallite. Marine evaporite deposits are widespread. In North America, for example, strata of marine evaporites underlie as much as 30 percent of the land area. Marine evaporites produce most of the salt that we use, the gypsum used for plaster and the potassium used in plant fertilizers.

Fig. 5.3: (a) Non-marine Borate deposits, Turkey (b) Gypsum beds thinly bedded

5.2.3 Chemical Precipitate Deposits

Chemical precipitate deposits (also known as chemical sedimentary deposits) are those deposits in which the ore minerals are precipitated from solution as a result of chemical or biochemical reactions in lake or sea water to form stratiform mineral deposits. The principal metals that occur in this type of deposit are; iron, manganese, copper, lead and zinc, as well as some non-metallic minerals such as carbonates and phosphates. The characteristics of chemical sedimentary deposits include their stratiform and syngenetic nature, the relatively fine grained and uniform texture of the ore minerals, the presence of sedimentary features, and close association with the non-metallic components, Chemical sedimentary mineral deposits of importance are:

(1) Sedimentary Iron deposits
(2) Sedimentary Manganese Deposits
(3) Sedimentary (Sediment Hosted) Copper deposits
(4) Sedimentary Uranium deposits
(5) Sedimentary Carbonate and Sulfate deposits

Sedimentary Iron Deposits

Sedimentary iron deposits are iron-rich sedimentary rocks that contain >15% iron. They are classified into three types, in decreasing order of importance: (a) banded iron formations, (b) sedimentary ironstones and (c) bog iron ores.

Banded Iron Formations (BIFs)

These are unique rocks of Precambrian age composed of alternating layers of iron-rich and silica-rich (chert) components, and a number of other iron bearing minerals. The BIFs are usually well-bedded or banded, elongated bodies that can extend for several kilometers. They are characterized by alternating layers of iron oxides and silica in the form of chert (Fi. 5.4). Hematite (Fe_2O_3), and magnetite (Fe_3O_4) constitute the major ore minerals, but iron carbonate (siderite), silicates (greenalite, minnesotaite, stilpnomelane), and sulfides (pyrite and very minor pyrrhotite) are also found. These mineral assemblages represent the oxide, carbonate, silicate and sulfide facies of iron formations respectively, although the oxide facies is the most predominant. BIFs usually contain about 15 to 40%Fe, but can be enriched by metamorphism, lateritic weathering and supergene enrichment. Economic deposits are considered as units with thicknesses of >30 meters, strike lengths of several kilometers and grades of >30% Fe.

BIFs are found in cratonic shield areas throughout the world (Labrador, Venezuela, Brazil, Russia, India, West Africa, South Africa, and Australia). Notable examples are the extensive deposits in the Lake Superior district of Minnesota-Wisconsin-Michigan in the U.S. and Ontario, Quebec and the Labrador-Quebec belts in adjoining areas of Canada, BIFs are also prominent in the Hamersley Basin of Western Australia, and several deposits in Brazil. India, and South Africa. In many parts of West Africa, BIFs can be found in Guinea, Liberia and Nigeria. There are three types of BIFs based on their characteristics, mode of occurrence and relative age: Lake Superior type, Algoma type and Rakuten type:

Lake Superior type BIF is the most common, and was formed in marine shelf sediments that are usually free of clastic material. The rhythmic banding of iron-rich and iron-poor chert layers is a prominent and distinctive feature that normally ranges in thickness from a centimeter to up to a meter, and can be traced for several kilometers (Fig. 5.4).

Fig. 5.4: Lake Superior type Banded iron formation

Algoma-type BIF is found in Archean high-grade terrains and closely associated with Archean greenstone belts emplaced in submarine volcanic settings. They are relatively small, usually less than 50m thick and found within volcanogenic sediments and rocks. They are similar to the Lake Superior type in appearance but they lack prominent and continuous banding of alternating iron-rich and silica rich bands (Fig. 5.5a). Large deposits of Algoma type BIF of middle Archean age occur in the Abitibi greenstone belt in Canada and the Guyana and Liberia shields in South America and West Africa respectively.

Rapitan-type BIF is found as small lenses in Late Precambrian rocks, and is the least important in terms of volume or size of occurrence. This type of BIF is closely related to glaciogenic marine sediments. Rapitan-type *BIF* is the least important in terms of volume of occurrence and size of deposits. The deposits are composed mainly of hematite and chert (Fig. 5.5b). Examples occur in the Yukon Territory, Canada, and the Bremer iron formation in South Australia.

Fig. 5.5: (a) Algoma type BIF (b) Rapitan type BIF

Sedimentary Ironstones

Sedimentary ironstone are low-grade, iron-impregnated sandstones, usually but not always characterized by oolitic and sometimes pisolitic textures. They can be distinguished from BIFs by their lack of banding and contain no primary colloidal silica (chert). The ironstones occur as reddish-brown earthy sandstone beds ranging in thickness from a few centimeters to tens of meters (Fig. 5.6a). In tropical environments, the iron-rich beds may be subject to lateritic weathering and supergene enrichment, and carry a lateritic cap several meters thick. Sedimentary ironstones are all Phanerozoic in age (compared to BIFs that are Precambrian).

The common ore minerals in ironstones are hematite, goethite and magnetite. Texturally, the ironstones appear as oolitic, pisolitic, or pelletal grains that may have been entirely or partially replaced by interlayered limonite and hematite. The oolitic texture (or oolite) consists of concentric rings of Fe-bearing minerals, namely hematite siderite and chamosite assemblage or goethite around sand grains or fragmental fossil or mineral material (Fig. 5.6b). There are two types of ironstones distinguished on the basis of mineralogical and mining attributes: the *Minette type* (limonitic) and *Clinton type* (hematitic).

Minette-type (named after Lorraine, France) ironstone is the most common and widespread, and is composed of oolites of siderite (Fe_2CO_3), *chamosite*, a siliceous iron mineral (($Fe_4Si_4O_{16}(OH)_8$) and goethite. The deposits were formed in shallow, near-shore marine environments and are found extensively in Europe, particularly in England, the Lorraine area of France, Belgium, and Luxembourg.

Clinton-type (named after an occurrence in New York, USA) deposits contain oolites of *hematite*, siderite, and chamosite, and are found mostly in the eastern Atlantic seaboard of the United States. The Cretaceous sedimentary ironstones in Nigeria (Agbaja and Enugu) belong to the Minette type, and are made up predominantly of goethite, magnetite and kaolinite but no chamosite.

The sedimentary ironstones differ from the Precambrian BIFs in the absence of chert, presence of oolitic textures and containing more alumina and phosphate. They are much smaller in size in both thickness and lateral extent. The iron concentrations in the oolitic ironstones are generally lower than in BIFs. Iron and phosphorus reach an average 34%Fe and 1%P_2O_5 respectively. The origin of sedimentary ironstones is well known. They were deposited in fluvial to shallow-water marine environments where they are associated with a variety of sedimentary rocks, including limestones, siltstones, shales and sandstones. Examples of these deposits are the Cretaceous Agbaja ironstones of central Nigeria, and the Minette- or Clinton-type ores that occur in Jurassic sediments in Lorraine and other areas of France and in the Silurian (Clinton Beds) of the United States respectively.

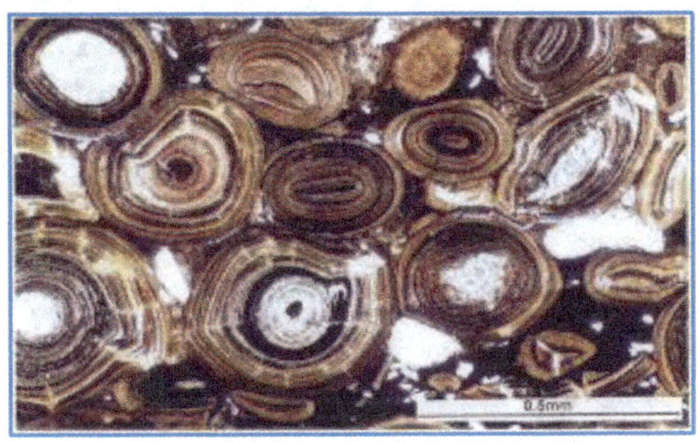

Fig. 5.6: (a) Sedimentary Ironstones, Nigeria (b) Oolitic-Pisolitic texture of Ironstone

Sedimentary Manganese Deposits

Sedimentary manganese deposits are quite similar in their mode of occurrence and spatial affinity with iron deposits. This relationship is attributed to the similarity in chemical behavior and how they are transported and concentrated in rocks. Sedimentary manganese deposits are formed by the precipitation of Mn from seawater. They commonly occur as laterally extensive sedimentary beds enriched in manganese in association with marine sediments, such as dolomite, limestone, chalk and black shale. The manganese beds can be >50 meters thick, and over 50 kilometers in lateral extent. Manganese ores often display a variety of textures, including oolites and pisolites, rhythmic laminations, slumped bedding, and abundant fossils. These deposits are the main source of manganese in the world. The dominant ore minerals are housemanite (Mn_3O_4), manganite ($Mn_2O_3.H_2O$), psilomelane, pyrolusite, rhodochrosite and other manganese minerals. Primary ore is commonly further enriched by the supergene process.

Manganese deposits can be classified into three types; (a) BIF-related manganese deposits which are mainly Proterozoic in age (Fig. 5.7a), (b) Manganiferous (oolitic) deposits (Fig. 5.7b), and (c) Manganese nodules (equivalent to bog iron ores). Most of the sedimentary manganese deposits of the world are Proterozoic in age, with minor occurrences in the Cenozoic. The largest deposits of the BIF-related manganese are located in the Kalahari region of South Africa, and in Brazil with many smaller deposits occurring in China. Cenozoic deposits which are mostly oolitic are centered around the Black Sea region of central Europe extending from Turkey to Bulgaria. Manganese nodules have been found on the deep seafloor as concretions formed by concentric layering of iron and manganese hydroxides around a core which may be rock or fossil fragments (Fig. 5.8a). The marine deposits are extensive but are currently not economic to extract.

The origin of manganese deposits is similar to that of iron ores. These deposits formed in shallow marine depositional environments (15-300 m deep) in lakes and seas, commonly in sheltered sites around islands, along some areas of continental shelf and the interior basins. Manganese precipitation is believed to take place in stratified water masses at the interface between anoxic seawater and near surface oxygenated waters (Fig. 5.8b). Extreme Fe fractionation is caused by a low solubility of iron in low Eh environments where Fe precipitates as iron sulfide. A subsequent increase in Eh and/or pH of Mn-rich water may produce Mn-rich, Fe-depleted chemical sediments. The sources of Mn include submarine volcanic activity and chemical weathering and transportation from continental to shallow marine environments.

Fig. 5.7: (a) BIF-hosted Manganese (b) Sedimentary Manganese Ore

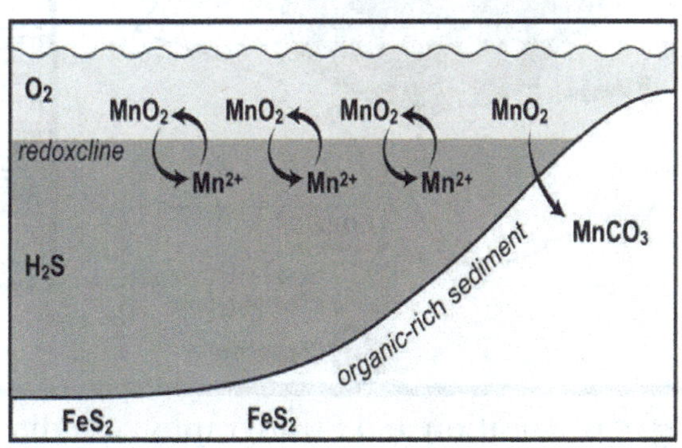

Fig. 5.8: (a) Manganese nodules (b) Manganese origin in sedimentary basins

5.2.3.1 Sediment Hosted Stratiform Copper Deposits

Sedimentary or sediment-hosted copper deposits are a significant group of copper deposits that occur within sedimentary rocks ranging from shales, siltstones to sandstones, and sometimes in carbonate. Sedimentary copper deposits are a large, diverse group of deposits that include some of the richest and largest copper deposits in the world. Twenty to twenty five percent of world copper production comes from this type of deposit which is also an important source of silver and cobalt, particularly the central Africa Copper belt of Zambia and Zaire which are the world's most important source of cobalt. Other classic examples are the Kuperschiefer copper deposits in north-central Europe, extending from Germany to Poland.

Copper mineralization occurs as finely disseminated, stratiform, sulfides of which chalcocite and bornite are the dominant copper minerals with lesser amounts of native Cu, chalcopyrite, galena, hematite and pyrite. Mineralization is usually continuous within beds or may follow old channels and in some places the mineralization may cut beds. Disseminated sulfides are usually stratiform along bedding planes. Bleaching of the sediments due to reduction usually occurs because mineralization is usually localized at reduction front between oxidizing (red sandstone) to reducing (green sandstone); Lateritic weathering produces reddish crusts while secondary copper minerals (azurite and malachite) form gossans due to supergene enrichment. Deposits may show mineral zoning with centers of chalcocite-bornite grading into outer zones of chalcopyrite-pyrite, and peripheral galena-sphalerite.

The origin of these copper deposits has been widely debated. The ore distributions indicate several generations of sulfides that are syngenetic, but extending into early diagenetic and late diagenetic processes. Sulfur isotopic data on sulfides suggest the derivation of sulfur essentially from the bacterial reduction of seawater sulfates. The mineralizing brines were generated from sea water in hypersaline lagoons during the deposition of the host rocks. Changes of Eh–pH and salinity probably were critical for concentrating copper–cobalt mineralization.

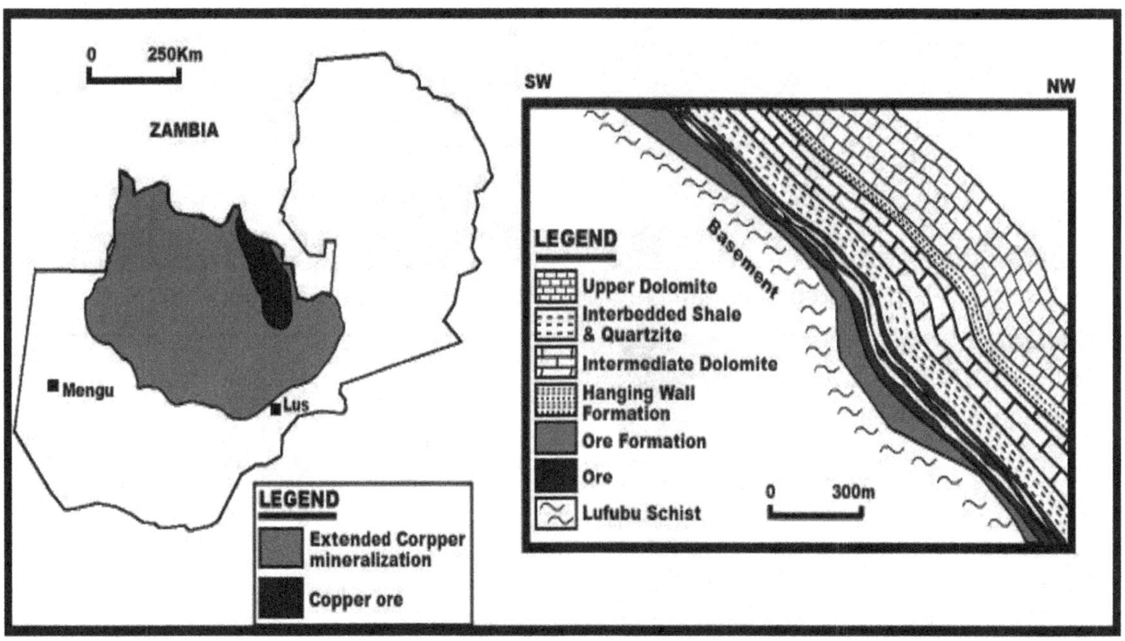

Fig. 5.9: Location and Stratigraphy of sedimentary copper deposits, Zambia

5.2.3.2 Sedimentary (Sandstone type) Uranium Deposits

Sediment-hosted uranium deposits consist of the minerals of uranium and sometimes vanadium that are found within clastic sedimentary rocks, mostly sandstones and some conglomerates that were deposited in a fluvial or marginal marine environment. Impermeable shale/mudstone units often occur immediately above and below the mineralized sandstone beds. Most of the deposits are tabular (stratiform) or channel fills. They constitute about 28% of world uranium proven reserves and 40% of inferred resources, and are of major economic importance in Kazakhstan, Uzbekistan, USA and Niger Republic who are among world's leading producers of uranium. Orebodies of this type are commonly low to medium grade (0.05-0.35%U) and individual orebodies are small to medium in size, with a maximum range of 50,000 tons.

Mineralization occurs as lenses, pods, layers, or concave "rolls" and irregular masses of ore that are grossly conformable with enclosing sediments (stratiform), but may sometimes show cross-cutting relationship in detailed observations (stratabound). There are three subtypes based on mode of occurrence: (a) tabular deposits, (b) channel fill deposits, and © roll front deposits. Most of the mineralization is concentrated within the intergranular spaces in the clastic host rocks, and may be associated with organic debris, or replacement of fossil logs. The dominant minerals in these deposits are uraninite (or pitchblende), and secondary uranium hydroxides such as coffinite [U(Sio4kiOH)4x], and vanadium mineral montroseite [VO(OH)]. Some sulfide minerals including pyrite, sphalerite and galena are sometimes found within the deposits. The amounts of uranium, (and vanadium) can vary considerably, both within and between deposits.

Examples of sedimentary uranium deposits occur in Niger, Gabon (Franceville Basin), and in the eastern part of Africa, in the Karoo Formation (Malawi, Tanzania, Zambia and South Africa), and in the United States in the Colorado Plateau area of Colorado, Navajo county in Arizona, Utah, New Mexico, the Powder Basin of Wyoming, the Gulf Coast Plain in south Texas and several deposits in China (Fig. 5.10). In the Niger Republic, uranium deposits occur within Carboniferous to Cretaceous sandstones as tabular deposits (with rare roll fronts) that are irregular, elongate or lenticular and parallel to the depositional trend of the host arenites of fluviatile to deltaic in origin. Mineralization mostly within fluvial paleo-channels and unconformities. The main ore minerals are pitchblende and coffinite. Niger accounts for about 6% of world uranium production.

The origin of sedimentary uranium deposits is controversial, but overwhelming evidence suggests that the ore deposition involved a combination of processes that range from syngenetic to diagenetic. Most of the tabular deposits were formed from uranium leached from weathered uranium-bearing source rocks by oxidizing solutions and precipitated on encountering reducing environments within permeable sandstones. Uranium exists in nature as tetravalent (U^{4+}) or hexavalent (U^{6+}) forms. The latter is soluble and can be easily carried by surface or subsurface water until it reaches a reducing environment where it deposits again in tetravalent form. The reductants varied at different deposits and included organic matter such as woody fragments, fossils, coalified wood, structureless organic matter, oil and natural gas. Most of the uranium was introduced during the deposition of the host sediments and concentrated during diagenesis by formation waters.

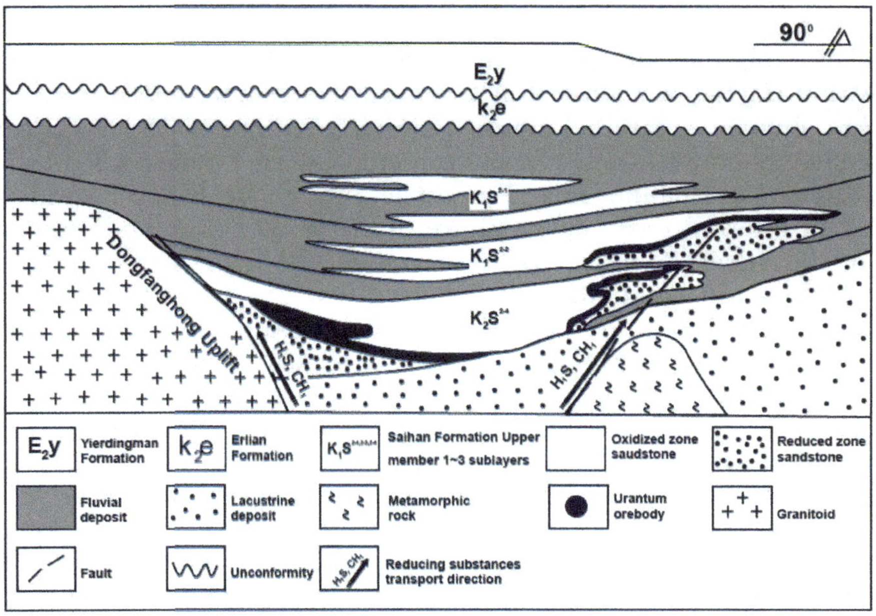

Fig. 5.10: Tabular uranium deposits in basal sandstones, China.

5.2.4 Biogenic Mineral Deposits

Biogenic sedimentary mineral deposits are those formed by the accumulation of the skeletal remains of organisms including plants, animals and microorganisms. Biogenic mineral deposits include non-metallic minerals and rocks, and solid mineral fuels These occur mainly as carbonates, siliceous deposits, phosphatic deposits, and plant organic remains as coals, tar sands and bituminous shales. (Fig. 5.11).

- Calcium carbonates - Limestone and Dolomite
- Siliceous deposits - Diatomite, Chert
- Phosphate deposits
- Plant remains (Coal, Lignite)
- Marine organics (Tar Sand, Oil Shale)
- Biogenic Sulfur

Carbonates and Siliceous Deposits

Limestone is an industrial rock deposit composed mainly of calcium carbonate ($CaCO_3$), usually in the form of calcite or aragonite. Dolomite is composed more of magnesian carbonate (MgCO3). Limestone is a sedimentary rock formed mostly from the biogenic precipitation from seawater of the skeletal remains of lime-secreting marine organisms and foraminifera. These deposits are major courses of calcium carbonate used for cement and lime production. Limestones are found in most sedimentary basins of the world. Diatomite is a very fine-grained siliceous sedimentary rock composed of the remains of fossilized diatoms. It is chemically inert and used as a filtration medium and a component in cement.

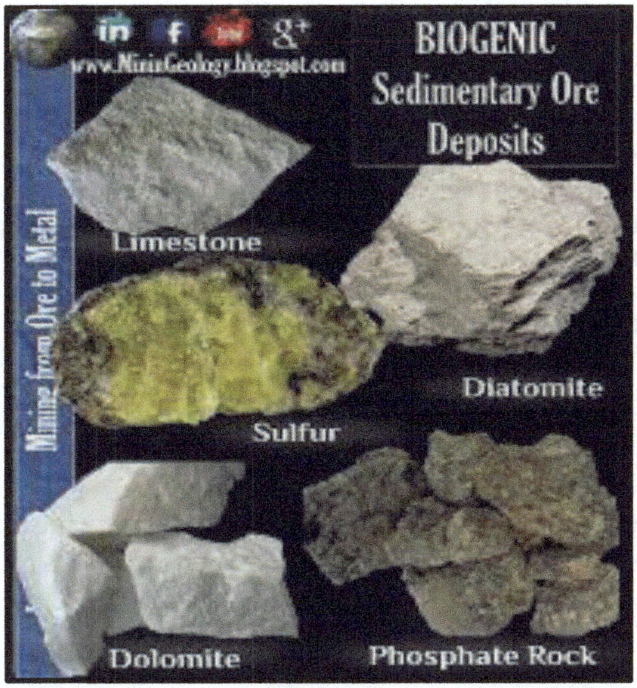

Fig. 5.11: Biogenic sedimentary ore samples

Phosphorite (Phosphate Rock)

Phosphorite, phosphate rock or rock phosphate are marine sedimentary rocks that contain high phosphate concentrations exceeding 4% P_2O_5, but usually in the range of 15–20% for commercial purposes. Phosphate is mostly contained as cryptocrystalline fluorapatite ($Ca_5(PO_4)_3F$) or as hydroxyapatite ($Ca_5(PO_4)_3OH$). Phosphorites are often associated with carbonate rocks, shales, and sandstones that occur in extensive layers, which may cover tens of thousands of square kilometers. Phosphorites usually show textures and structures similar to fine-grained limestones and sometimes may represent diagenetic replacements of existing carbonate minerals by phosphates. They also can be composed of pisolites, ooids, fossils, and clasts that are made up of apatite. About 2% of world phosphate is derived from guano - the excreta of birds.

The origin of phosphorites is related to the precipitation of phosphorus in a variety of depositional environments that are very shallow, near-shore marine or low energy environments. The phosphate accumulation is derived from organisms. Currently, areas of oceanic upwelling cause the formation of phosphates because of the constant stream of phosphate brought from the large, deep ocean reservoir. This cycle allows continuous growth of organisms that utilize phosphate for shells. Significant deposits of phosphorites in the United States include the Phosphoria Formation in Idaho, and the Monterey Formation in California. Major deposits also occur in Peru, as well as North Africa in (Morocco and Mauritania) and Togo in West Africa.

Coal Deposits

Coal is a black or brown combustible sedimentary rock formed from plant remains of ancient vegetation which has been consolidated between other rock strata and transformed by the combined effects of microbial action, pressure and heat over a considerable time period. This process is commonly called 'coalification'. Coal is classified into four types based on the degree of coalification (carbon content) or the rank of coal: (i) Anthracite (86-97% C), (ii) Bituminous Coal (45-85% C), Sub-Bituminous Coal (35-45% C) and Lignite or Brown Coal (25-35% C). Coal occurs as layers or seams, ranging in thickness from millimeters to many tens of meters. Coal is composed mostly of carbon (50–98 per cent), hydrogen (3–13 per cent) and oxygen, and smaller amounts of nitrogen, sulfur and other elements. It also contains water and particles of other inorganic matter. When burnt, coal releases energy as heat which has a variety of uses. Coal is broadly separated into brown and black which have different thermal properties and uses.

Brown coal (lignite) has a low energy and high ash content, and is used to generate electricity in power stations located at or near mine installations. Black coal is harder than brown coal and has a higher energy content. Thermal (steaming) coal is black coal and is used mainly for generating electricity in power stations where it is pulverized and burnt to heat steam generating boilers. Metallurgical (coking) coal is black coal with low ash suitable for making coke, which is used in the metallurgical production of pig iron. These coals must also have low sulfur and phosphorus contents, and are relatively scarce and attract a higher price than thermal coals. Coal is formed when dead plant matter decays into peat and is converted into coal by the heat and pressure of deep burial over millions of years. Coal deposits are widespread in most continents. South Africa has substantial resources of coal.

Biogenic Sulfur Deposits

Sulfur deposits are composed of native sulfur that form extensive beds within evaporite-bearing sedimentary basins. Note that major sources of non-biogenic sulfur are associated with hot spring and volcanic regions. Native sulfur is formed when hydrocarbons come in contact with sulfate minerals in presence of liquid water. Most economically important elemental sulfur deposits being mined today are biogenic replacements of anhydrite or gypsum. These occur in two ways: in cap rocks over salt diapirs and in stratabound deposits. They are formed in evaporite basins where petroleum-bearing beds underlie anhydrite or gypsum and occur where joints or faults permit water, hydrocarbons, and sulfate-reducing bacteria to rise into the evaporites. The bacteria oxidize hydrocarbons to CO2, reduce sulfate ions to H2S, and alter gypsum to calcite. The H2S is converted to polysulfides and these are oxidized to elemental sulfur by CO2 in anaerobic environments. The main use of sulfur is in the preparation of chemical compounds, but other applications include fireworks, matches, fumigant and fertilizer. Substantial biogenic sulfur deposits are found in the United States Gulf Coast in Texas and Louisiana, and southeastern Mexico.

Review Questions

1. Define sedimentary mineral deposits? Briefly describe the various types. Cite examples.
2. Describe mode of occurrence and mineralogy of sediment-hosted copper deposits, and explain their genesis.
3. Explain the differences between alluvial and marine placer deposits. What are the metallic ores commonly associated with these deposits?
4. Banded iron formations are the world's leading sources of iron ore. Describe the various types and explain how they are different from sedimentary ironstones.
5. Write short notes on the following: (a) coal, (b) biogenic sulfur, and (c) phosphorite. Briefly explain their mode of origin.

Chapter 6: Metamorphism-Related Mineral Deposits

6.1 Introduction

Mineral deposits formed by metamorphic processes are the concentrations or enrichment of economic minerals due to the effects of heat, pressure, fluids and dynamic movements. During regional and contact metamorphism, temperature, pressure and fluids are the dominant agents, and their effects on pre-existing rocks can lead to the accumulation of useful minerals. Few metalliferous ores owe their existence to regional metamorphism, but countless massive sulfide and oxide ores have been significantly altered or enriched by metamorphic effects.

The impacts of high temperature, pressure and tectonic deformation associated with regional metamorphism during orogenic events can affect bodies of existing mineral deposits leading to their deformation and acquisition of features typical of metamorphic rocks, including recrystallization of minerals in solid state and general coarsening of grain size. Other changes include replacement of low-density minerals by minerals of high volumetric mass; hydrous minerals are replaced by anhydrous ones, and amorphous matter crystallizes into new minerals. High grade metamorphism of granitic rocks can also lead to partial melting and production of anatectic pegmatites that can contain economic minerals. All these transformations can occasionally lead to mineralization or the enrichment of pre-existing low-grade ores. For example, the conversion of minerals such as hematite (Fe_2O_3) to specular hematite (xFe_2O_3) or magnetite (Fe_3O_4) enhances ore grade.

Metamorphic ore-forming fluids are commonly produced by dehydration reactions at deep crustal levels, and the heating, devolatilization and remobilization of connate waters in sedimentary sequences. These heated waters could be responsible for the formation of certain types of deposits, including gold, lead and zinc. This group of deposits could be classified as hydrothermal by some authors, but as metamorphic by others. Fluids derived from igneous intrusions along their contact with sedimentary host rocks produce contact metasomatic deposits. Thus, mineral deposits related to metamorphic processes (metamorphism-related) are grouped into four types.

1. Metamorphic Deposits
    - Industrial Minerals (Graphite, Asbestos)
    - Gemstones (Garnets, Zircon)
2. Metamorphosed Deposits
    - Iron Ores (Ferruginous Quartzites)
    - Manganese Ores (Gondite)

3. *Metamorphogenic Deposits
    - Orogenic gold
4. Contact Metasomatic Deposits
    - Skarn (e.g. tungsten, copper)

*The term "metamorphogenic" is used after Walter L. Pohl (2011).

6.2 Metamorphic Deposits

Metamorphic deposits are those mineral deposits formed directly by metamorphic effects of heat and pressure. Not many metallic deposits are created solely by metamorphism, but several industrial minerals and gemstones come to mind because they are formed during regional metamorphism of various rocks including mafic/ultramafic, pelitic and aluminous sediments. Such minerals include graphite, garnet, asbestos, talc, corundum, kyanite, sillimanite, and gemstones (e.g. ruby, sapphire, and emerald). Industrial rocks such as marble, slate, granite gneisses and quartzites are of considerable economic value in building construction.

Graphite is formed from metamorphism of coal, while talc and asbestos deposits are formed by regional metamorphism of ultramafic rocks. Several examples are found within the Precambrian greenstone and schist belts. Kyanite and sillimanite are formed from the metamorphism of aluminous sediments and shales. Marbles used as decorative stones are produced by regional metamorphism of limestones. Gemstones formed by regional metamorphism include varieties of garnets (spessartine, rhodonite, almandine, and grossular), corundum and spinel.

Unusual occurrences of metamorphic pegmatites formed probably by anatexis of local crustal material and containing mineralization have been reported from high grade metamorphic terrains such as the Lewisian Gneiss Complex, Scotland. Metamorphic pegmatites can be formed by partial melts of mobile constituents that concentrate during high grade metamorphic differentiation. This is in contrast to igneous pegmatites that are formed from residual fluids of magmatic fractionation. Migmatites are examples of rocks created when a metamorphic rock partially melts at high temperature and produces a feldspar-rich fraction similar to pegmatite (Fig. 6.1). Economic mineral deposits may be associated with this process. A prominent example possibly is the Rossing uranium deposit in Namibia where pegmatite and alaskite within high-grade metamorphic terrain may have produced the uranium deposit.

Figure 6.1: Migmatite containing anatectic feldspar-rich layers

6.2.1 Graphite Deposits

Graphite deposits are examples of a truly metamorphic deposit because the economic mineral was created during regional metamorphism of carbonaceous rocks. Graphite is a crystalline form of alaskites which are three major types; (a) flake (crystalline) graphite; (b) amorphous (microcrystalline) graphite, and (c) vein (lump) graphite. Crystalline flake graphite deposits consist of flakes of graphite that occur in a variety of rocks including marble, paragneiss, iron formation, quartzite, and pegmatite and, in extremely rare cases, serpentinized ultramafic rocks Amorphous (or microcrystalline) graphite deposits occur as intercalated lenses and beds within metamorphosed siliceous sediments such as quartz-mica schists, micaceous quartzites, or micaceous feldspathic quartzites, and rarely within marbles. They consist mainly of small graphite particles intergrown with other mineral impurities. Examples are found in Alabama and Texas, USA, where they are more than 100 feet thick; they range in length from a few feet to as much as half a mile. Amorphous graphite deposits.

Vein-type graphite deposits are the most economically important group that occurs as fissure-filling veins, pods and lenses, irregular bodies, stockworks, and saddle reef within metasedimentary rocks in the upper amphibolite and granulite facies grade of metamorphism. They are also found in zones associated with contact metamorphism by igneous intrusions. Vein graphite is composed of coarsely crystalline graphite-rich fragments that are typically 0.5 to 0.8 cm in diameter and are commercially described as "lumps". The origin of the vein graphite deposits is controversial, but they are believed to be products of high-grade granulite facies metamorphism in the presence of a CO_2 rich fluid under low fO_2 conditions. Examples include extensive deposits in Sri Lanka, where graphite veins are currently mined.

6.3 Metamorphosed Deposits

Metamorphosed mineral deposits are those deposits that were originally sedimentary rocks but subjected to metamorphism which may have produced changes such as enhanced metal content and increased grain size. Such changes often increase the economic value of such deposits. Examples are ferruginous quartzites where hematite with lower iron content is converted to magnetite. In some banded iron formations (BIFs) in Archean terrains, they retain most of the "banding" and a lot of their primary sedimentary features, but undergo some textural and mineralogical changes with increased crystallinity and more magnetite ore. They are called *"itabirite"* in Brazil. Itabirites have been described from Guyana and West Africa, and also occur in the Hamersley province of Australia

6.3.1 Ferruginous Quartzites

Ferruginous quartzites are usually found in medium to high grade metamorphic terrain in which iron-rich bands containing magnetite and/or hematite alternate with crystalline bands of quartz, with minor sillimanite. They are the products of regional metamorphism of iron-rich sandstones. Unlike the banded iron formations (BIFs), ferruginous quartzites do not contain chert or show consistent banding, although they do contain quartz derived from metamorphism of quartz grains. The ferruginous quartzites are not always banded, and if they are, the bands are not consistent in either width or lateral continuity because they are products of metamorphic differentiation and not sedimentary bedding. The development of banding increases with metamorphic grade similar to those observed in banded gneisses. The iron minerals are hematite, usually specularite and magnetite. Ferruginous quartzites are at times described as *hematite-magnetite quartzites*. Examples of iron ore deposits of ferruginous quartzites occur widely in Precambrian shield areas of the world, and include several occurrences in Indi, Brazil, Russia and Ukraine. In central Nigeria near Okene, Kogi State, several deposits of ferruginous quartzites extending for several kilometers are intercalated within banded gneisses and amphibolites. The most extensive deposits of ferruginous quartzites in the world are located in the Krivoi Rog basin (Fig. 6.2a), and the Kursk magnetic anomaly in eastern Ukraine. Grades of iron ores range from 25 to 45% Fe.

Fig. 6.2: (a) Ferruginous quartzite, Ukraine (b) Metamorphosed Manganese Ore. Japan

6.3.2 Metamorphosed Manganese Deposits

Metamorphosed manganese deposits are formed by the metamorphism of sedimentary manganese ores and manganiferous sediments, sometimes described as gondite. The grade of metamorphism can range from low-grade greenschist facies to high-grade granulite facies. The mineralogical assemblages include manganese oxides and manganese silicates. The texture of the manganese ores is enhanced by coarser and granoblastic texture as the grade of metamorphism increases. Examples of these metamorphosed manganese ores are those found in India (Andhra Pradesh), and Namibia. The Mn occurs in three stratigraphic zones associated with ferruginous quartzites and schists. Inter-banded oxide and silicate layers are represented by mineral assemblages: braunite-jacobsite-hausmannite, and spessartite-rhodonite, diopside-acmite-quartz respectively. The impact of metamorphism is to increase the manganese oxide content. Metamorphosed manganese deposits occur in metamorphosed volcanic sediments at the Shiromaru mine near Tokyo, (Fig. 6.2b).

6.4 Metamorphogenic Deposits

This class of metamorphism-related deposits was proposed by the German School of thought on economic geology led by W.L Pohl. These deposits are purported to have been formed from fluids generated during prograde metamorphism of pelitic rocks by collecting water from the dehydration and devolatilization of sedimentary minerals. Such fluids definitely can be mobilized and produce lateral secretion of metals that are concentrated as mineral deposits. Such fluids are arguably different from the connate or sedimentary pore fluid that could also be heated and mobilized during metamorphism.

According to the theory, two end member models have been proposed to define a range of possibilities occurring in nature to generate the mineralizing fluid: (a) the prograde devolatilization model, and (b) the retrograde leaching model. In the first case, the mineralizing fluids are thought to have been generated and to have acquired their solute content within source rocks undergoing prograde metamorphism. Then, they flow towards lower pressure zones, possibly accompanied by further reactions with country rocks. Finally, changes in the chemical or physical environment along the flow conduits potentially induce mineral precipitation. Retrograde leaching, on the other hand, occurs after peak metamorphism, when a still-hot metamorphic complex is flooded and cooled by waters from outside; this will normally happen during uplift and extensional tectonics, and the resulting convection cells are open to the surface.

Metamorphogenic deposits may not be easily recognized as such in the field, where they appear as epigenetic hydrothermal mineralization that may be difficult to differentiate from hydrothermal deposits related to metamorphic fluids (described earlier in this chapter). Only the application of a broad spectrum of laboratory methods, such as isotope geochemistry and fluid inclusion study can allow a correct genetic diagnosis. A good example commonly cited are the orogenic *gold deposits* found in metamorphosed Precambrian greenstone-schist belts in the form of structurally controlled quartz and quartz-carbonate veins.

6.4.1 Orogenic Gold Deposits

Orogenic gold deposits are characterized by "mesothermal" quartz or quartz-carbonate veins in deformed and metamorphosed terranes of Precambrian greenstone belts and schist belts. They are described as "orogenic" because of their association with the deformation of orogenic belts or convergent plate margin. The metamorphic grade of the host rocks ranges from greenschist facies to upper amphibolite facies or rarely to granulite facies. The gold-bearing veins are localized within major structural features spatially associated with deep-seated crustal deformation zones. Hydrothermal alteration surrounding the veins contains the same minerals as in the gangue – quartz, carbonates and sulfides with additional low-temperature silicate minerals such as sericite, albite, biotite, and chlorite. Well-known deposits of orogenic gold are found in the Archean greenstone belts of Australia and Canada, and the Proterozoic Birimian greenstone belts of the West African Craton in Ghana, Guinea, Mali and Burkina Faso. Orogenic gold constitutes a very substantial source of gold worldwide. The origin of the orogenic gold deposits is attributed to metamorphic fluids derived from intra-basin sources or underlying continental crust. Ores were deposited from low-salinity, near-neutral, and CO2 rich fluids driven through the crustal sequence of metavolcanic and metasedimentary rocks, where they leach out gold transported as chloride or sulfide complexes. The fluids were channeled along major structural discontinuities where reaction with wall rocks resulted in the deposition of gold.

6.5 Contact Metamorphic/Metasomatic Deposits

Contact metamorphic minerals are those formed solely by reaction between host rock and hot igneous intrusion. without fluid interactions. Except for some industrial minerals (e.g. graphite), not many economic minerals are formed solely by this process. The presence of fluids emanating from the hot igneous intrusion (metasomatism) produces more economic deposits. Contact metasomatic or skarn deposits are formed where a felsic magma (granite/granodiorite) intrudes sedimentary rocks, particularly carbonate rocks, resulting in the deposition of new metallic minerals with a unique assemblage of calc-silicate gangue minerals. This process is commonly accompanied by the discharge of acidic fluids from the intrusive body which facilitates chemical reaction with the sedimentary rocks and the ore deposition.

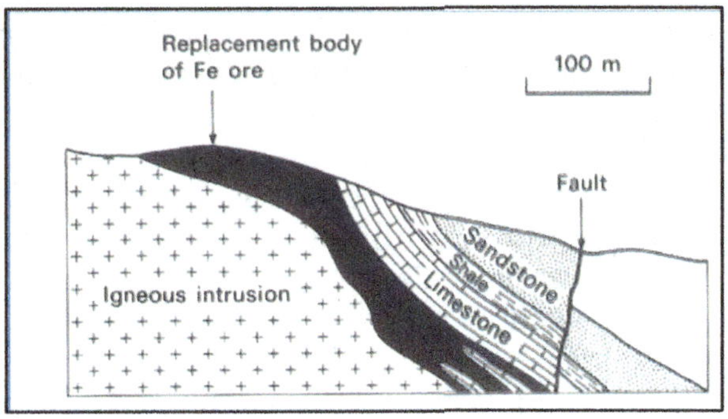

Figure 6.3: (a) Contact metasomatic deposit (b) Skarn copper ore

6.5.1 Skarn Deposits

Skarn deposits are replacement deposits because they replace the carbonate host rocks during the chemical reactions that lead to their deposition (Fig. 6.3a). Common ore minerals in skarn deposits include both sulfides and oxides, such as molybdenite, chalcopyrite, covellite, (Fig. 6.3b), sphalerite, galena, and oxides: cassiterite, wolframite, magnetite, scheelite, gold and hematite. Gangue minerals in skarns are unique, and consist of quartz, various garnets, amphiboles, pyroxenes, calc-silicates, olivine, talc, epidote, garnets and anhydrite. Skarn deposits are important as sources of iron, molybdenum, gold, tungsten, copper, lead, zinc, and tin. They constitute the world's premier sources of tungsten (more than 70% of world's tungsten production).

Review Questions

1. Classify metamorphism related mineral deposits and describe their major characteristics, citing examples.
2. Discuss the various factors involved in the formation of mineral deposits. What are the potential sources of metals in these deposits?
3. Describe how metamorphism can enrich the value or grade of existing mineral deposits. Cite examples.
4. What are skarn deposits. Give examples, and describe their characteristics and origin.
5. Write short notes on the following: (a) metamorphogenic deposits, (b) formation of graphite, and (c) ferruginous quartzites.

Chapter 7: Surficial Mineral Deposits

7.1 Introduction

Surficial mineral deposits are the concentrations of useful minerals near or at the Earth's surface as a result of surficial geological processes such as chemical weathering, and supergene enrichment. *Chemical or lateritic weathering* involves oxidation and hydrolysis and the dissolution or leaching of soluble rock constituents (usually Si and Fe-Mg), accompanied by the precipitation or concentration of a less soluble residue which may include
Fe hydroxides in laterites. The textures of the parent materials are usually obliterated in these residues by repeated solution and deposition of secondary minerals. In contrast, *supergene enrichment* involves deposition of secondary minerals by the action of descending rain water or groundwater, and typically occurs well below the water table. Enrichment often preserves most of the textures of the parent material, until other processes such as metamorphism or lateritization may subsequently intervene in the process. Two groups of deposits are formed by surficial processes: residual and supergene.

The differences between products of *chemical/lateritic weathering* and *supergene enrichment* are noteworthy. Lateritic deposits are concentrated by the leaching away of unwanted material and residual concentration of useful materials without any external contribution. In supergene enrichment, useful materials are removed and concentrated elsewhere, usually at depth or across a lateral front. Thus, a weathered profile may contain both lateritic and supergene deposits of a specific metal or material. Mineral deposits formed by surficial processes can be classified into three types:

(1) Residual and Lateritic deposits - Clays, Bauxite, Lateritic Iron, Lateritic Manganese, Lateritic Nickel and Lateritic Gold.
(2) Supergene Enrichment deposits - Supergene Copper/Sulfides, Supergene Manganese, Supergene Niobium and Supergene Gold.
(3) Infiltration deposits - Roll front Uranium, Roll front Molybdenum and Karstic Bauxite.

7.2 Residual and Lateritic Deposits

Residual deposits are accumulations of valuable minerals by chemical weathering after the undesirable constituents of rocks have been removed. The term 'lateritic deposit" is used to describe residual deposits in tropical and subtropical regions where the residual minerals contain iron oxides and hydroxides. The major requirements for the formation of residual deposits are: (a) parent rock of appropriate composition, (b) leaching of the undesirable constituents by intense chemical weathering and adequate drainage, and (c) low relief and crustal stability so that the residual accumulation is not eroded away as soon as it is formed. Residual deposits accumulate more in tropical and sub-tropical environments because of the effects of high precipitation and temperatures that accelerate intense weathering.

Aluminum and iron are the two most common elements concentrated by lateritic chemical weathering. After soluble elements are removed by hydrolysis, Fe and Al usually form hydroxides that are more stable under surficial conditions. Other transition metals such as Ni, Mn and occasionally, native metals like gold, may also be concentrated. Accumulation of these elements near the surface often produces lateritic ores. Common examples of residual deposits produced by weathering are: (1) bauxites, (2) lateritic iron ore, (3) lateritic nickel, (4) lateritic manganese, and 5) lateritic gold.

7.2.1 Bauxite Deposits

Bauxite is the main ore of aluminum, and is found mainly as a residual deposit formed by intense chemical weathering of many different Al-rich parent rocks, including granitic rocks, syenites, phonolites, shales, pelitic schist etc. Clay minerals commonly represent intermediate stages, and bauxite can grade into clay or laterite. Bauxite ore consists of the minerals, *gibbsite* (Al (OH)3, *boehmite* (γ-Al (OH) 3) and *diaspore* (AlO (OH), and are sometimes mixed with kaolinite with minor quartz and iron hydroxides (Fig. 7.1a).

Lateritic bauxites are found mainly in tropical and subtropical regions where iron hydroxides are precipitated along with bauxite minerals. They generally occur as blanket deposits formed in the upper parts of thick surface lateritic deposits that form a cap over plateaus in tropical areas for example, in Guinea, Ghana, Guyana and India). Other types of bauxite not related to weathering are the sedimentary and karst varieties. Most of the world's aluminum is mined from lateritic bauxite deposits. The world's largest producer of bauxite is Australia's Weipa Mine in northern Queensland (Fig. 7.1b), while other countries with large reserves include Guinea, Ghana, Brazil, Vietnam, and Jamaica.

The origin of lateritic bauxites is closely related to intense tropical weathering of aluminous parent rocks under conditions of high temperature and high rainfall which produce leaching of silicates, iron oxides and kaolinite, and the precipitation of Al-hydroxides, usually gibbsite. The chemical changes associated with formation of lateritic bauxites include the removal of silica, alkalis, calcium, magnesium, iron and titanium, and concentration of alumina. The formation of bauxites has been attributed to a two-stage process: (a) chemical weathering and leaching of the parent rock to produce aluminum and iron-rich residue; and (b) further leaching of iron oxides and kaolinite to produce pure Al-hydroxides (gibbsite). Sometimes the bauxite is derived directly from underlying kaolinite (Fig. 7.1b). The formation of lateritic bauxites is determined by the following conditions: (1) favorable parent rock with minerals that are soluble. (2) effective rock porosity that facilitates movement of water, (3) High rainfall with intermittent dry periods, (4) tropical warm climate, (5) low to moderate topographic relief, (6) good drainage, and (7) prolonged stability without much disturbance.

Fig. 7.1: (a) Bauxite sample (b)Lateritic bauxite overlying kaolinite, Australia

7.2.2 Lateritic Iron Ores

Lateritic iron ores are formed by intense chemical weathering involving leaching and concentration of residual elements in ferruginous host rocks to form economic mineral deposits. Suitable parent rocks are ferruginous sandstones, ultramafic rocks and low-grade banded iron formations (itabirites). The enrichment process involves the simple removal (dissolution) of silica and alumina in the primary ore accompanied by textural destruction, oxidation and the precipitation of residual iron oxides usually as soft or friable ore with enriched iron content. The ore grade of these deposits is in the range of 40 to 60% Fe depending on the original concentration in the parent material.

Lateritic weathering of various parent materials has been widely studied. In the case of low-grade banded iron formations known in Brazil as "itabirites", lateritic weathering under wet tropical conditions has resulted in well-developed weathering mantle, with a specific succession of horizons of different chemical, mineralogical and textural compositions capped by a ferruginous duricrust. These deposits can be up to 30 meters thick with goethite and hematite as main ore minerals. A typical profile consists from top to bottom: (a) ferruginous cap (canga); (b) soft hematite of mostly hematite and goethite and no quartz, (c) weathered itabirite (friable and disaggregated), (d) slightly altered itabirite (minor changes) and (e) fresh itabirite (hard itabirite with hematite/magnetite) (Fig. 7.2a).

In Brazil, India and Australia, the lateritic weathering of itabirites (low grade iron-rich rocks) in wet tropical conditions has resulted in a well-developed weathering mantle, with a specific succession of horizons of different chemical, mineralogical and textural compositions capped by a ferruginous duricrust (Fig. 7.2b). These deposits can be up to 20 meters thick. The lack of alumina in the itabirite and the intense leaching of silica inhibited the formation of clay minerals. The dominant iron minerals are goethite and hematite. This progressive lateritic modification tends to produce a varied surface complex, including massive, pisolitic, and botryoidal, or nodular aluminous/siliceous goethite and hematite, forming ferricrete or duricrust. Examples of these deposits exist in Brazil, Guyana, Cuba and Malaysia.

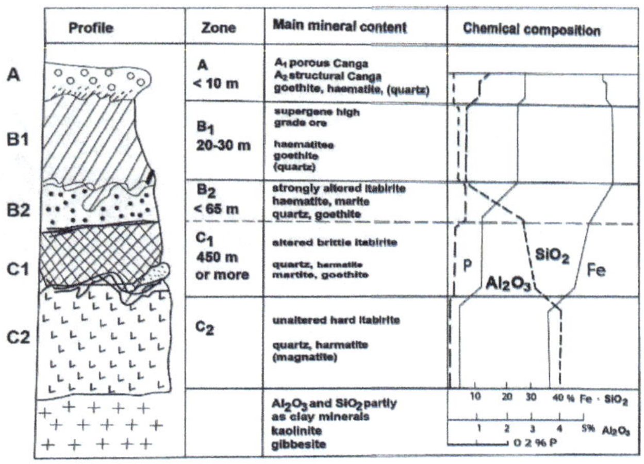

Fig. 7.2: (a)Typical lateritic iron ore profile, Brazil (b) Lateritic Iron Ore, India

7.2,3 Lateritic Nickel Deposits

Lateritic nickel ores account for over 60% of global nickel supply. They are formed from the intensive chemical weathering of olivine-rich ultramafic rocks such as dunite, peridotite and serpentinite (Fig. 7.3a). The lateritization process removes the most soluble elements magnesium (Mg), calcium (Ca), and silicon (Si), and concentrates the least soluble elements (iron (Fe), nickel (Ni, manganese (Mn), and chromium (Cr). Mechanical weathering of the host material along fractures and faults increases the surface area exposed to the water driving the chemical weathering. This initial nickel content less than 0.1- 0.2% in the parent material may be enriched ten times in the course of lateritic weathering.

Two kinds of lateritic nickel ores have been distinguished: (a) *limonite types* and (b) *silicate types*. A third type which contains clay minerals (clay type) has been identified in some localities. Limonite-type lateritic ores (or oxide type) are highly enriched in iron due to very strong leaching of magnesium and silica. They consist largely of goethite and contain 1-2% nickel incorporated in goethite. The silicate type (or saprolite type) nickel ore formed beneath the limonite zone. It contains generally 1.5-2.5% Ni and consists largely of Mg-depleted serpentine in which nickel is incorporated. In pockets and fissures of the serpentinite rock, green garnierite, a magnesium-based silicate, can be present in minor quantities, but with high nickel contents of mostly 20-40%.

Typical lateritic nickel ore deposits are very large tonnage, low-grade deposits located close to the surface. They are typically in the range of 20 million tons and upwards at a grade of 1% Ni with some examples approaching a billion tons of ore. The ore deposits are restricted to the weathering mantle above the parent rock and tend to be tabular, flat and y large, covering many square kilometers of the Earth's surface. The dominant minerals are silicates of nickel, such as garnierite (Fig. 7.3b). Examples of lateritic nickel deposits are found in Western Australia, Cuba, New Caledonia, Oregon, USA, Greece and USSR.

Figure 7.3: (a) Laterite Nickel Profile (b) Laterite Nickel Ore with Garnierite

7.2.4 Lateritic Gold Deposits

Lateritic gold deposits are a relatively recent discovery, but they have acquired some importance, especially in Western Australia where large nuggets of gold have been found in laterites (Fig. 3.62). These deposits are formed by weathering of bedrock containing some fine gold, and chemical enrichment of gold occurs at the base of the lateritic weathering profiles. The source rocks are disseminated gold deposits or weathered ultramafic rocks or greenstones that contain gold specks. The gold is often deposited within iron oxides and/or aluminous concretions or hard crusts (Fig.7.4).

Examples of lateritic gold deposits can be found in Western Australia, Guyana, India and Brazil. In the Carajas region of Brazil, native gold is found within iron oxide-hydroxide nodules in the ferruginous zone where gold orebodies have been delineated. The Boddington bauxite mine in Western Australia was until its recent closure, the largest gold producer in Australia with annual production of 2400 kg. Sources of the gold at Boddington mine are quartz veins and hydrothermally altered rocks related to the formation of the bauxite.

7.2.5 Lateritic Manganese Deposits

Lateritic weathering of manganese bearing metamorphic rocks (gondite) have produced some deposits in Brazil and many parts of Africa (Ghana, Ivory Coast, and South Africa). These concentrations within the lateritic profile consist of several Mn oxyhydroxides including manganite in association with Fe hydroxides. Lateritic manganese ores are often high-grade and of superior quality. Source rocks for lateritic Mn include manganese quartzites, carbonates and volcano-sedimentary rocks that contain exhalative-sedimentary Mn- enriched horizons.

Figure 7.4: Lateritic Gold in Surficial Deposits, Australia

7.3 Supergene Deposits

Supergene mineral deposits are those deposits formed by secondary enrichment near the earth's surface. Supergene processes which form this type of deposit are a remarkable form of weathering that leads to secondary enrichment of sulfide and oxide ores through the action of descending surface waters. This enrichment usually takes place by the alteration of primary sulfide or oxide minerals from the oxidized zone above the water table and the accumulation of the dissolved metals in the enriched zone below the water table. Supergene enrichment can occur in many types of ore deposits, but it is notably important in three categories of deposits: 1) sulfide deposits; 2) sedimentary Fe and Mn formations; and; 3) carbonatite Nb, Th and REE deposits.

7.3.1 Supergene Copper Deposits

Porphyry copper deposits are low grade disseminated mineralization of copper sulfides and pyrite in porphyritic granitic rocks. Secondary enrichment of porphyry copper deposits in the southwestern United States, Mexico, Peru, and Chile is an important factor in making those deposits economically viable ores._Enrichment blankets at many porphyry copper deposits form large laterally extensive zones (up to 30-40 square miles) that typically range from 200 to 1,000 feet in thickness, but thin laterally to 10 to 100 feet. The enrichment blanket at the Chuquicamata deposit in Chile is reported to attain thicknesses of up to 2,500 feet. Enriched sulfide ores are mainly composed of sooty to fine-grained chalcocite that coats and/or replaces pyrite and copper-bearing sulfides. (Fig. 7.5).

7.3.2 Supergene Iron Deposits

Many of the world's richest iron ores have undergone supergene enrichment of banded iron formations through the action of descending meteoric waters where the grade of iron ores has been increased two to three times. Prominent examples can be found in Western Australia and Brazil. The leaching process involves the removal of silica in the BIFs much more than the remobilization of Fe. The removal of silica produces more compact iron ores and the conversion of hematite to magnetite which enriches the Fe. In the supergene iron ores of Carajas, Brazil where primary itabirites with about 35% Fe have been enriched to up to 68% Fe. *Canga*, a cap rock consisting of detrital iron-rich fragments, cemented by limonite, shows an average grade of about 62% Fe. Zones enriched by lateritic weathering may merge with supergene enriched zone

7.3.3 Supergene Manganese Ores

Supergene manganese deposits are formed where chemical reactions taking place within tens of meters from the surface redistribute and concentrate manganese at lower depths within the weathering profile overlying manganiferous sediments and carbonates (Fig. 3.64). Sedimentary manganese deposits, especially those in carbonate rocks or with volcanic affiliation have been secondarily enriched to become ores by this process involving the leaching of primary Mn carbonates or silicates and conversion to Mn oxides such as pyrolusite and manganite, usually by oxidation and/or dissolution followed by migration and re-precipitation. A good example is the sedimentary manganese deposit at Groote Eylandt in Northern Australia where supergene enrichment has increased the ore grade from 15% to about 50%. Other examples include deposits in Brazil and Ghana where supergene ores occur at depth within the lateritic profile (Fig. 7.5b). Most common minerals found on the deposit are a variety of Mn oxides, pyrolusite, cryptomelane and psilomelane, with minor quartz, kaolinite, goethite and hematite. These phases are a combination of both primary and/or supergene minerals.

7.3.4 Supergene Niobium Deposits

Supergene niobium deposits occur in laterites formed from intense chemical weathering of carbonatites and secondary enrichment of disseminated niobium minerals. The process involves the decomposition of pyrochlore, apatite and magnetite, followed by removal of soluble carbonates and formation of secondary minerals, commonly Ba, Sr, K or Pb-bearing pyrochlore. The supergene enrichment of Nb occurs by the formation of Nb-bearing rutile, brookite, and anatase. The St. Honore carbonatite in Brazil hosts the largest niobium deposit in the world in which supergene enrichment occurs within a lateritic profile of up to 300 m deep. The Nb contents of the supergene deposits can be enriched up to 10 times or more above those of the primary carbonatite. Most primary deposits range in concentration from 0.1 to 0.3%Nb compared to enriched values averaging 3% Nb. Examples are found mainly in Brazil (St. Honore, Araxa, Catalao, and Bingo) which accounts for about 92% of the global production of Nb.

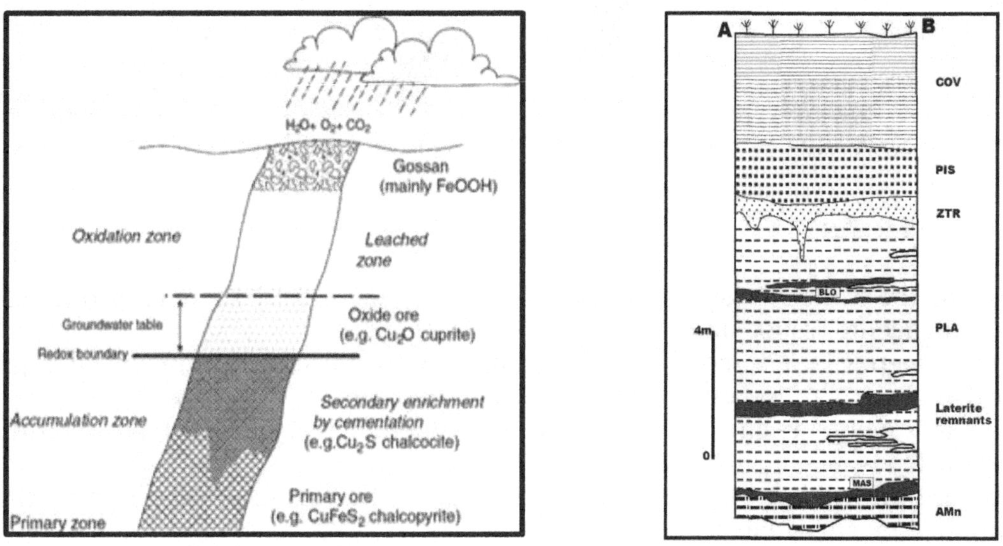

Figure 7.5 (a) Supergene Copper deposit (b) Supergene Manganese enrichment (black)

7.3.5 Infiltration Deposits

Infiltration mineral deposits are concentrations of minerals formed by the consolidation of readily soluble elements by downward percolating waters in the supergene zone of sedimentary rocks. Mineral formation occurs when oxygenated meteoric water leaches and transports elements by infiltration through permeable strata such as sandstones precipitating ores from solution due to reducing conditions (Fig. 7.6a). The so-called "roll front" uranium deposits are the most typical example of infiltration deposits. They are usually found in porous and permeable sandstones and conglomerates. Roll fronts mark the redox boundary in space and time where the infiltrating meteoric water carrying uranium lost its capacity to oxidize the percolated sandstone, and was forced to precipitate uranium from solution along a reducing front (Fig. 7.6b).

Uranium occurs in two valence states, the reduced form U4+ and the oxidized form U6+. The uranyl ion U6+ is highly soluble in oxidized fluids but as groundwater infiltrates slowly through the aquifer, it becomes unstable when it encounters a reducing agent such as organic matter, and will precipitate as U4+ in the form of uranium minerals, coffinite and uraninite, with occasional vanadium-rich clays and minor sulfides. Roll front deposits are found extensively in the United States, in Colorado, Utah, Wyoming and Texas. Other metals such as molybdenum, copper, iron, vanadium, silver, and Pb-Z-Ba-F could form roll front deposits. Some typical characteristics of roll front deposits include (a) appearance as crescent shaped bodies; (b) the convex side points to the direction of hydraulic gradient; © most orebodies consist of several interconnected cells, and (d) individual roll front deposits are quite small, but when combined can extend for considerable distances.

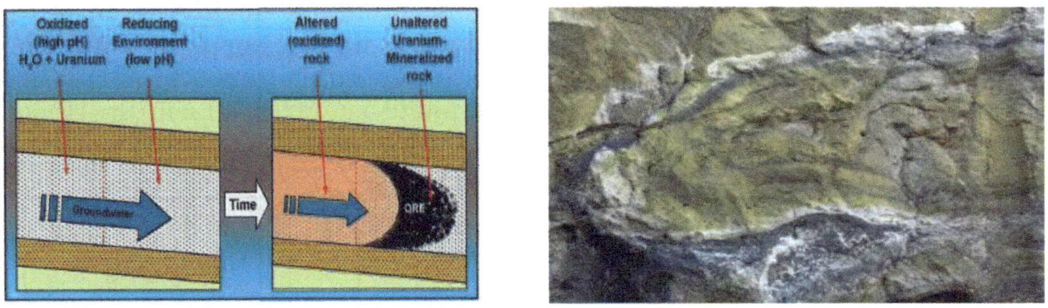

Figure 7.6: (a) Diagram of Infiltration Deposit (b) Roll front Uranium in Dakota Sandstone

Review Questions

1. Describe the three categories of deposits that are formed by surficial processes. What are the main agents responsible for their formation?
2. Lateritic deposits are mostly found within the tropical and subtropical regions of the world. explain why and give examples of such deposits.
3. What is supergene enrichment? How do supergene enriched deposits different from lateritic deposits.
4. What are infiltration deposits? With the help of illustration, describe how the roll front uranium deposits are formed.
5. Write short notes on (a) lateritic gold deposits, (b supergene niobium deposits and (c) bauxite deposits. Explain the differences in ore-forming processes.

Chapter 8: Industrial Minerals and Rocks

8.1 Introduction

Industrial minerals and rocks are earth materials that do not contain metals or fuel that are used in construction, and industrial applications. They are valued not for their metal content or energy source but mainly for their physical properties, and sometimes chemical composition. Most of these minerals (and rocks) are widely used for building and construction, and in various manufacturing industries to produce a variety of products. Examples include limestone, mica, barite, gypsum, marble, phosphate, salt, diamond, granite, talc and soapstone. Other industrial minerals are valued as gems for the jewelry industry. In recent years, the interest in industrial minerals (and rocks) has increased considerably due to the substantial contribution of this class of mineral resources to the economy of many developed and developing countries. In the advanced countries, industrial minerals provide inputs in many industrial processes, while in some developing countries with little industrial infrastructure, significant portions of their foreign exchange derive from exports of industrial minerals. The rise in the exploitation of industrial minerals and rocks as a result of their widespread occurrence and uses has galvanized the efforts of geological exploration for these deposits, as part of a concerted strategy to enhance the solid minerals economy of many developing countries. A good example is Nigeria where over 80% of its annual solid mineral production is attributable to industrial minerals and rocks, particularly limestone which is used for cement production, a material in high demand for building construction, and granite which is used as crushed stone aggregates mainly for road construction. Compared to metallic minerals, industrial minerals show the following differences:

- Do not contain metals.
- Do not show shiny surfaces or luster typical of metals
- Valued for physical and chemical properties and not metal content
- Good insulators rather than good conductors of electricity and heat
- Lack malleability and ductility, and break or fracture easily when hit hard
- Found predominantly in sedimentary rocks
- More abundant than metallic minerals

Industrial minerals and rocks can be subdivided into four groups on the basis of their composition and uses:

1. Industrial minerals (e.g. barite, feldspar, graphite, salt, talc)
2. Industrial rocks (e.g. phosphate rock, diatomite, limestone, nepheline syenite)

3. Construction materials (e.g. sand, gravel, crushed stone, laterite)
4. Gemstones (e.g. diamond, emerald, sapphire, aquamarine)

8.2 Types of Industrial Minerals and Rocks

8.2.1 Industrial Minerals

Industrial minerals are a group of non-metallic minerals that are used as raw materials for industrial applications. Industrial minerals can also be defined more specifically as "geological materials that are not sources of metals, fuel, or gemstones, but are mined for their commercial value". Industrial minerals are valuable, not for the metals they contain, but for their physical and chemical properties. They are used mostly in their natural state or after minimal beneficiation, either as raw materials or as additives in a wide range of household and industrial products. A list of common industrial minerals is shown in Table 8.1.

Table 8.1: Types of Non-Metallic Minerals and Rocks

Industrial Minerals		Industrial Rocks	Construction Materials	Gemstones
Andalusite	Kyanite	Carbonatite	Aggregate	Amethyst
Asbestos	Mica	Diatomite	Building Stone	Aquamarine
Barite	Olivine	Limestone	Cement	Diamond
Bentonite	Magnesite	Nepheline Syenite	Brick Clay	Emerald
Borates	Potash	Phosphorite	Gravel	Garnet
Clays	Quartz /Quartz Sand	Basalt	Laterite	Ruby
Corundum	Sillimanite	Marble	Pumice & Perlite	Sapphire
Feldspar	Talc	Granite	Quartzite	Topaz
Fluorspar	Trona (Soda Ash)	Serpentine	Sand	Tourmaline
Graphite	Wollastonite	Soapstone	Sandstone	Zircon
Gypsum	Zeolites		Slate	
Halite	Zircon/Zircon Sand		Quartzite	
Kaolin				

The end uses of industrial minerals are varied. Some are used as sources of important chemicals (e.g. halite for sodium chloride *and* borax for borates). Some are used for the production of building materials (e.g. gypsum for plaster and kaolin for ceramics). Others are applied as agricultural fertilizers or animal feed (e.g. apatite for phosphate and sylvite for potassium). Still, others are utilized in many other ways, such as abrasives (e.g. diamond and corundum) or fillers (e.g. mica, kyanite). Many industrial minerals have a low unit value (low cost per ton of material) which makes them less valuable except close to consumers. It is often said that "*Without a market, an industrial mineral is merely a geological curiosity*".

8.2.2 Industrial Rocks

Industrial rocks are indurated aggregates of minerals that are used as raw materials for the manufacture of industrial products or for ornamental applications without any mineral separation. This definition distinguishes *industrial rocks* from *industrial minerals*. Apart from the metal mining industry, the ornamental stone industry is the oldest known mineral industry that used rocks for decorative purposes and not as construction material. Many geologists have sometimes described *construction materials* such as gravel, sand and clay as industrial rocks. However, because these materials are loose and not indurated, or compact, they are not rocks *in sensu stricto*. The best-known industrial rocks are limestone, phosphate, coal, diatomite, carbonatite and nepheline syenite, and ornamental stones -marble, granite, serpentinite and soapstone.

8.2.3 Construction Materials

Construction materials are a group of unconsolidated or loose earth materials and crushed or cut rocks that are used directly in building and road construction. Examples include sand and gravel, crushed rock aggregates, laterite, brick clay, building stones such as slate and sandstone, and cement materials.

8.2.4 Gemstones

Gemstones (or gems) are a special class of industrial minerals used in the jewelry industry. A gemstone is defined as a mineral or rock, which can be used in jewelry after cutting or faceting and polishing. Gemstones often occur as well-formed crystals of colorful minerals that are spectacular objects of beauty when cut and polished for use as jewelry or personal adornments. Gems are so valuable in their rarity, and sufficiently durable to give lasting pleasure to their owners. Although almost all gemstones are minerals, there are a few rocks (such as lapis lazuli) and organic materials that are not minerals (such as amber, jet, and pearl) but are also used for jewelry, and are therefore often described as gemstones.

8.3 Characteristics of Industrial Minerals

Industrial minerals possess some unique physical and chemical characteristics that distinguish them from other earth materials. Industrial minerals are:

1. Utilized because of their physical characteristics and/or chemical composition rather than their metal content.
2. Used in industrial applications with very little or no beneficiation.
3. Most industrial minerals have high *place value* but low *unit value*. (Place value reflects the importance of the value of the product in relation to its market while unit value indicates the real market value of a unit of the product e.g. one ton of sand)
4. Economic importance depends on the availability of markets, market location, and transportation costs. Without a market, an industrial mineral deposit is merely a geological curiosity.
5. Variable functions in industry. (Products derived from industrial minerals are interrelated in various industrial processes (e.g. use of barite and bentonite in drilling mud).
6. A single material may form the basis for a wide range of applications in different industries (e.g. limestone is used for cement, production of lime, road aggregates etc.).

8.4 Uses of Industrial Minerals

Industrial minerals are widely used in the following major areas of industrial activity: Table 8.2is a list of industrial minerals and their uses.

a) ***Building Industry***: Several industrial minerals are used in the production of building materials. Gypsum is used in the production of drywall, cement plaster and additive for hardening of Portland cement. Andalusite, kyanite and sillimanite are used as additives in ceramic tiles for walls and floors, Asbestos is used in roofing and ceiling tiles and for insulation. Clays, feldspar and quartz are important raw materials in the manufacture of floor and wall tiles and structural clay products.

b) ***Manufacturing Industry***: Several products are manufactured from industrial minerals. In the manufacture of sheet glass and containers, quartz and silica sand, with additives of fluorspar, borates and soda ash are used. For ceramics, a variety of industrial clays is combined with feldspar as flux in clay bodies and glaze. For abrasives, minerals such as diamond, zircon, garnet, corundum are used. For filler manufacturing, ground calcite and zeolites are used as a filler in paint, paper and plastics. Kaolin is used in coating of paper and as paint fillers and medicinal products.

c) ***Agriculture Industry***: Most fertilizers are produced from natural minerals and rock products such as apatite (phosphates), nitrates, potash, salt, and sulfur. Calcite, and to a lesser degree dolomite, when pulverized (ground to a fine powder), may be used to condition clayey soils and neutralize acid soils in order to promote plant growth.

d) ***Chemical Industry***: Industrial minerals are used in the production of several chemicals. Such minerals include limestone (for lime), fluorite, sulfur, halite (salt), nitrates and borates.

e) ***Metallurgical and Refractory Industry:*** The metallurgical industry uses industrial minerals as fluxes in the processing of iron ores and other metals. Limestone (for lime), borax, silica and dolomite are metallurgical fluxes; magnesite, fire clay, graphite, bauxite, zircon, silica, and dolomite are used in furnace linings. Minerals commonly used for refractories include magnesite, bauxite, graphite, kaolin, and zirconium.

f) ***Jewelry Industry***: The jewelry industry is involved in the making, distribution, and repair of jewelry such as watches, rings, and necklaces using gemstones etc.

Table 8.2: Industrial Minerals and their Uses

Industrial Minerals	**Uses**	**Industrial Minerals**	**Uses**
Andalusite, Kyanite, Sillimanite	Refractories, ceramics and alumina bricks	**Kaolin**	Used in paper coating: for ceramic whitewares and pigment in paint
Asbestos	Roof and floor tiles, fire and heat resistant products.	**Mica**	Use as insulators, lenses of optical instruments
Barite	Weighting material in drilling mu; and filler or extender.	**Quartz / Quartz Sand**	Glass, ceramics and chemical apparatus
Bentonite	Component of drilling mud for deep wells	**Salt (Halite)**	Table salt, food seasoning and chemicals
Borates	Insulators, fiberglass and specialty glass, textile	**Talc**	Cosmetic powder, personal care products; filler in paints & ceramics

Clays	Ceramics, refractories and structural clay products	**Trona (Soda Ash)**	Flux for glass making, water purification; removal of pollutants
Corundum	Used for abrasives and alumina refractories and bricks	**Wollastonite**	Use in ceramics and friction products; filler in paints
Feldspar	Flux in ceramics. Fillers and extenders in paints and plastics	**Zeolites**	Cement additive; use as adsorbent, ion exchangers. Water filtration.
Fluorspar	Flux for melting of glass and enamels. Used for chemicals.	**Zircon**	Refractory bricks and liners; foundry casting applications.
Graphite	Used in lead pencils, lubricants		
Gypsum	Manufacture of drywall, cement and plaster of Paris.		

8.5 Importance of Industrial Minerals and Rocks

Industrial minerals constitute the most widely exploited mineral resource in both developed and less developed countries. In developed countries such as the United States, the annual dollar value of industrial mineral production has since surpassed that for metals, and continues to grow rapidly. In 2018, the value of non-metallic minerals and rocks produced in the United States was valued at $53 billion (US dollars) compared to the value of metals production worth $30 billion (USD). Industrial minerals are of great economic value as they serve as major raw materials in the manufacture of several industrial and daily-need products such as glass, abrasive, paper, chemicals, detergents, ceramics, fertilizers, and metallurgical and agricultural products. A wide range of everyday domestic and vital industrial products are made from industrial minerals. According to the Industrial Minerals Association of North America, the average American resident consumes about 24 tons of industrial minerals each year.

While in developed countries industrial minerals provide inputs in many industrial processes, it is a little different in some less developed countries with little industrial infrastructure, where significant portions of their foreign exchange derive from exports of industrial minerals and rocks. Also, the fact that industrial minerals (and rocks) are found almost everywhere (widespread occurrence) makes these materials good sources of income for local suppliers and communities. For example, sand and gravel are available in most local communities in Africa where they are excavated and used for building construction which generate income for local entrepreneurs and transporters. Industrial minerals are therefore vital to a modern economy – they underpin the manufacturing industry, construction and agriculture – they also have important environmental applications by conversion of mine wastes to beneficial use.

Owing to the scarcity of resources and supply issues, the European Commission on Raw Materials, recently designated some industrial minerals as *strategic minerals* that are "critical" to industries in Western Europe, e.g. graphite, fluorspar, phosphate, borates etc. Nigeria has designated limestone as a strategic mineral for national development. For developing countries such as Nigeria, the prospects for selling industrial minerals on the export market has been very encouraging in recent years, and this has led to increased mining activity by artisans. Moreover, the desire to locally manufacture products that use industrial minerals is seen as a mechanism to conserve foreign exchange reserves by reducing the importation of such products. The growth in the demand for industrial minerals can be attributed to the combination of high and increasing per capita consumption of goods produced from industrial minerals, as well as increasing income levels in developing countries.

For example, the consumption of cement and ceramic products has grown exponentially due to the booming building construction in African countries. Several household goods and life conveniences, such as automobiles, and solar panels require the use of industrial minerals. For example, the Dangote Group, a Nigerian multinational industrial conglomerate, founded by the billionaire businessman Alike Dangote. has invested significantly in the production of cement by using locally available industrial minerals and rocks. Currently, the company owns Dangote Cement LLC which is the largest cement production company in Africa, with a market capitalization of almost US$14 billion and has subsidiaries with cement factories in Benin, Cameroon, Ghana, Nigeria, South Africa and Zambia

The widespread exploitation of industrial minerals has produced positive economic impacts on many communities in developing countries, in large part due to the multiplier effect in which a single mineral in one production process often involves the use of several others. For example, the production of glass from quartz sands may require the use of soda ash, calcite, dolomite, feldspar, borax, gypsum and fluorspar. Above all, these minerals and rocks provide the raw materials for infrastructure development in which large volumes of sand, gravel, clay, crushed and dimension stone are consumed locally. Industrial minerals are therefore vital to a modern economy where they form the bedrock for a fast-growing manufacturing industry, construction and agricultural activities

8.6 Description of Selected Industrial Minerals

8.6.1 Barite

Barite (or Baryte) ($BaSO_4$) is a sulfate mineral and the main ore of barium. The mineral has a wide range of colors and crystal forms, but is commonly colorless, reddish-brown to white. Barite is heavy, with a specific gravity varying from 3.5 to 4.5 (Fig. 8.1a). The largest proportion of barite production is used in the petroleum industry as a weighting material to make drilling mud heavy and capable of lifting drill cuttings. Other uses of barite include as filler or extender, and additive to cement, plastic, rubber, and urethane foam; and plastic automobile paint primer; friction products (brake and clutch pads) and as a white pigment in paints. Global production of barite in 2019 was 9.5 million tons (Mt), and the leading producers were China (2.9 Mt), India (2.2 Mt), Kazakhstan (0.62, Iran (0.49), and Mexico (0.42 Mt).

The geological occurrence of barite is varied. Commercial deposits are of five types: (i) *contact metasomatic deposits,* (ii) *vein deposits*, (iii) *sedimentary (bedded)* deposits, (iv) *volcanic bedded deposits,* and (v) *residual deposits.* South Africa and Nigeria are the main producers of barite in Africa. Other countries are Morocco, Algeria and Egypt. In South Africa, barite occurs as metasomatic replacement deposits in Namaqualand, and as stratabound barite associated with sedimentary exhalative (Sedex) type massive sulfides in the Northern Cape Province. In Nigeria, barite occurs as veins and cavity filling deposits closely associated with Pb -Zn hydrothermal veins within the Cretaceous sediments of the Benue Trough in Cross River. Benue, Nasarawa, Taraba and Plateau states. Barite veins have also been found in the basement complex adjacent to the sedimentary basins.

Fig. 8.1: (a) Barite (b) Bentonite mine

8.6.2 Bentonite

Bentonite is a naturally occurring, very fine-grained and plastic material composed mainly of the clay mineral, montmorillonite. It is found mainly in areas with altered volcanic materials (Fig. 8.1b) or in sedimentary basins. Bentonite has a high swelling capacity and viscosity which are essential properties for use as a drilling mud component. There is also an increasing demand for bentonite for use in other areas, such as binders for iron pellets, cosmetics, pharmaceuticals etc. Generally, bentonites that are required for these applications are pure white bentonite with high viscosity. Bentonite is a clay of montmorillonite composition - a hydrous aluminum silicate containing small amounts of alkali and alkaline earth metals. For industrial purposes, two main types of bentonite exist: sodium and calcium bentonite. The sodium bentonite is the "true" bentonite which is very plastic with high swelling propensity. The Ca-bentonite has more calcium than sodium and less swelling capacity. However, the Ca-bentonite can be beneficiated by adding Na from Na2CO3 which improves its swelling capacity but not as much as the natural Na-bentonite.

Bentonite is used in drilling fluids to lubricate and cool the cutting tools, to remove cuttings, and to help prevent oil well blowouts. Much of bentonite's usefulness in the drilling and geotechnical industry comes from its unique rheological properties. Bentonite usually forms from the weathering of volcanic ash, or the devitrification and chemical alteration of some igneous material, most often in the presence of water. This allows for the enrichment in sodium that enhances the swelling capacity of bentonite. Bentonitic clays and calcium bentonite occur within sedimentary environments where they are associated with marine clay deposits.

The world leading producers of bentonite are the USA, China, Greece, India and Turkey. The largest producer in Africa is Morocco. Most Nigerian "bentonites" are of the calcium variety and not suitable for use except when beneficiated which can raise the cost of production. There are also a group of clays and shales that are sometimes described as "bentonitic clays", because they have certain properties similar to bentonite but are mostly Ca-montmorillonite with a small amount of Na. Nigeria is a major oil producer and the drilling activities associated with the development of petroleum and groundwater resources require the use of bentonite for drilling mud. Nigeria imports most of its bentonite in quantities exceeding 70,000 tons per annum.

8.6.3 Clays and Kaolin

Clays are naturally occurring, fine-grained materials containing one or more clay that become plastic when wet. Clays are important industrial minerals widely used in construction and manufacturing of ceramics, bricks, hydraulic cement, structural clay products, fillers and extenders, drilling mud, fiberglass, paper coating, furnace lining etc. Clays are usually classified as either *primary* or *secondary* depending on their geological history. Primary clays are found in the same place as the parent material from which they originated, while secondary clays have been transported by wind, water or both from their places of formation. Primary clays are either products of chemical weathering (residual clays) or hydrothermal alteration (hydrothermal clays). Secondary clays produced initially by weathering have been transported by water, wind, and ice and deposited in locations distant from the source material. The following are the various types of secondary clays: namely, *alluvial clays* (river deposition), *marine clays* (sea deposition) and *glacial clays* (ice action).

The main industrial properties of clay minerals include plasticity, porosity, firing color, tensile strength and vitrification (fusibility). All these properties are important in evaluating the potential industrial applications of clays. *Plasticity* is the most outstanding property that makes clays easily workable. *Porosity* is a necessary property that clay must have so that clay products can dry without cracking. *Firing characteristics* refer to the behavior and appearance (color) of clay products after being "fired" (subjected to high temperatures) at above 900° C. They are very relevant to clay's aesthetic value. *Vitrification* is the process of becoming glass-like when fired, which is a property related to its ability to fuse at appropriate temperature (fusibility). Although clay products never become absolutely vitrified or glass like, it is necessary that the clay becomes hard (or almost vitrified) at a reasonable temperature, usually 1200 to1500°C.

Clays are important industrial minerals widely used in building construction and manufacturing of ceramics, bricks, hydraulic cement, structural clay products, fillers and extenders, and additives in drilling mud, fiberglass, paper coating, furnace lining etc. Industrial clays can be classified into the following groups based on their industrial properties and usage.

- *China Clay/Kaolin* (primarily kaolinite with minor quartz, mica)
- *Ball Clay* (primarily kaolinite with illite, chlorite, smectite minerals, quartz)
- *Fire Clay* (kaolinite, halloysite, diaspore)
- *Common Clay* (illite, kaolinite and chlorite, others)
- *Bentonitic Clay* (smectite with feldspars, biotite, quartz)

Kaolin

Kaolin is the "king" of clays, and the purest of all clays and a very fine, soft white clay made up of more than 80% kaolinite (Fig. 8.2a). Highest quality kaolin contains almost pure kaolinite with >90% of particles less than 0.002 mm in size and of low to moderate plasticity. Minor amounts of mica and quartz may be present. Kaolin is formed mostly by the hydrothermal alteration of granitic rocks and pegmatites, and also by chemical weathering of several feldspathic rock types. Kaolin has some unique physical properties that are useful in industrial applications. These properties include shape, particle size, color, gloss, viscosity, softness and non-abrasiveness. Kaolin has long been used in the ceramic industry, especially in the production of china wares, fine porcelain and other table wares. However, the primary use of kaolin is in the paper industry where it is applied as paper coating for brightness and glossy appearance. Other industrial applications include use as a pigment in white paint and a filler in fiberglass, rubber, plastics, and cable insulators. Kaolin deposits can be evaluated for their brightness, particle size distribution, viscosity (important for the valuable paper-coating clays), the iron and potash contents, fired brightness, casting behavior and other ceramic properties. The quality of kaolin can be improved by additional processing and purification.

Kaolin is a special clay with a high unit value and is a commodity traded around the world. In 2019, a little over 28 million tons were produced worldwide with a market value exceeding $4 billion US dollars Most of the demand is from paper, ceramics, fiberglass, paints, rubber, plastics end-use industries, The world-leading producers are the United States (7 Mt), Uzbekistan (5 Mt), Czech Republic (3.8 Mt), Germany (3.8 Mt), and Korea (2.9 Mt). Nigeria produces some kaolin and kaolinitic clays that are widespread within areas underlain by the crystalline rocks of the Precambrian basement complex in central, southwest and northwest regions. Relatively pure kaolin is found in a few places such as the well-known Kankara Clay in Katsina State and Potter Smith near Jos in Plateau State.

Ball Clay

Ball clay is a white burning, but usually buff to gray-colored, very plastic clay that burns to a light color. It is a member of the kaolinite family that is renowned for its plasticity and green strength, and as such, it forms a major part of many ceramic applications. Ball clay consists of varying proportions of kaolinite, mica and quartz, with small amounts of organic matter and other minerals. The very fine-grained texture of ball clay enhances its plasticity and commercial value because it is added to kaolin to increase the workability and strength of various ceramic bodies, as well as burn white or near white when fired. Ball clays are in general, different from kaolin by their higher plasticity and much finer-grained texture.

Ball clay is used in many different industries, but in particular, forms a vital component in ceramic manufacturing because it provides the plastic body for these products. Ball clay is used in the production of sanitary ware, tableware and floor/wall tiles. A 'ceramic body' for sanitary ware typically includes 30% ball clay to provide plasticity and workability, 20% kaolin, 30% feldspar and 20% quartz/silica. Ceramic tableware and sanitary ware utilize ball clay to provide high plasticity and good white-fired color, combined with kaolin, feldspar and quartz. In wall and floor tiles, ball clay is added for its plasticity and bonding properties combined with talc, feldspar, quartz/silica and kaolin. Ball clay is also used in the production of glazes which are coatings for ceramic products to ensure a perfect finish.

The origin of ball clay is obvious by its common association with other rocks of sedimentary origin. They are usually formed in a marine environment (marine clay). As a result of its origin, raw ball clay may have a wide range of off-white colors but is most valued by the ceramics industry for its white-firing properties, Ball clays are not very common, but fortunately they are found abundantly in most of Nigeria's sedimentary basins, including in Ondo, Edo, Anambra and Delta States.

Fire (Refractory) Clay

Fire Clay is a clay capable of resisting high degrees of heat at temperatures of up to 1700°C without damage. To be referred to as a "fire clay", the clay material must be refractory, i.e. being able to withstand a minimum temperature of 1515°C without melting. Fire clays are also described as refractory clays. Fire clays may be plastic or non-plastic and composed of kaolinite with some mica or illite and some quartz and organic matter. Fire clays vary widely in physical properties as long as they are refractory. Some fire clays are very plastic (*plastic fire clay*), while some are stiff and lack plasticity (*flint fire clay*), with a range in between. The fired color may vary, but most fire clays burn white. Chemically, fire clays are characterized by relatively high alumina content, with a Si/Al ratio of about 1:2. Such refractoriness or resistance to heat means that the clay is relatively pure and free from iron. Fire clays are used for a variety of products, mainly in the manufacture of fire bricks and other refractory parts for kilns, furnaces, boilers, and melting pots. The origin of fire clays ranges from residual to sedimentary, but most are of sedimentary origin. In Nigeria, extensive fire clays occur in the Anambra Basin where they lie beneath the coal beds of Enugu in Enugu State, and in the East Dahomey Basin at Onibode Clay in Ogun State.

Common Clays

The term "common clays" was devised by the US Geological Survey to describe a type of earthy clay materials, including soil clays, lateritic clays, weathered shales and glacial clays that are plastic when wet and composed of some clay minerals, other silicates and organic matter. These clays are fine grained and typically exhibit plastic behavior when wet and can be molded into many desired shapes, dried and fired to produce clay products with a rock-like hardness. The common minerals are kaolinite, and/or illite and/or chlorite, sometimes with minor montmorillonite, quartz, feldspar, and calcite. In the raw form, common clays are red, brown, greenish, or gray, as a result of the presence of iron oxide and some organic matter. They are the principal raw materials used in the production of traditional pottery wares, construction bricks, roofing tiles, drain tiles and other structural clay products such as drainage or sewage pipes and decorative bricks.

Fig. 8.3: (a) Kaolin deposit (b) Fluorspar

8.6.4 Fluorspar

Fluorspar or fluorite (CaF2), is a mineral with a vitreous luster, perfect octahedral cleavage, and relative softness (hardness of 4). It comes in a variety of colors: colorless, white, yellow, blue, green, rose, red, brown or nearly black, but the purplish variety is the most common (Fig. 58.2b). Ideally, fluorspar contains 51.1% calcium and 48.9% fluorine. For industrial applications, fluorspar is used for its chemical composition as a fluxing agent for smelting in the steel industry, and as additive in the production of decorative glassware and enamels. The purest grades of fluorite are a source of important industrial chemicals, hydrogen fluoride (HF) and hydrofluoric acid (70% HF), which are the intermediate source of most fluorine-containing fine chemicals, such as CFCs (chlorofluorocarbons), HFCs and HCFCs. These are used as industrial solvents, refrigerants, aerosol propellants and plastic foams. Fluorspar is found in a variety of geological settings but economic concentrations are mainly hydrothermal and sedimentary.

8.6.5 Gypsum

Gypsum is a soft mineral made up of hydrated calcium sulfate (CaSo4.2H2O). It is often white or colorless when pure, but can be tinted grayish, reddish, yellowish, or brownish due to impurities of SiO_2, Al_2O_3, MgO, K_2O and Na_2O. Texturally, gypsum can appear in three forms: *massive* (alabaster), *fibrous* (satin spar), or in *transparent crystals* (selenite). Gypsum is a very versatile mineral with varied applications in industry due to its softness (2 on Mohs' scale), and being moderately water soluble at low temperatures. Nearly all gypsum is used in three prime applications: (i) building construction (plaster, plasterboard, etc.) involving 33% of the market; (ii) cement (where it is used as a setting retarder): 61% of the market; and (iii) agriculture (mostly for soil conditioner and fertilizero.

Gypsum is one of the more common minerals in sedimentary evaporite basins. It is an evaporite mineral that occurs as massive or layered beds in sedimentary deposits in association with other evaporites such as halite, anhydrite, sylvite and other salts or carbonates. It originated from the evaporation of sea or lake water, and precipitation of the minerals. Gypsum can also be found as veins in argillaceous sediments. In 2019, the world's largest producer of gypsum from mines was the United States, with a production of 20 million tons (Mt), followed by Iran and China, both producing 16 Mt. African countries that produce gypsum are Morocco, Mauritania, Egypt, South Africa, and Tanzania which is reported to have one the largest gypsum deposits in the world that are yet to be developed due to lack of infrastructure.

The demand for gypsum has increased in Africa in recent years due to the increased manufacture of cement for building and road construction. Gypsum represents 5% of cement components. Nigeria alone has about thirteen cement factories with an annual production of over 40 million tons of cement. Although Nigeria still imports gypsum to satisfy demand, some gypsum occurrences are being mined especially in the northeastern region of Borno, Yobe, Gombe, and Adamawa States where they are intercalated within the Cretaceous to Cenozoic sediments of the ancient Lake Chad basin

8.6.6 Halite (Salt)

Halite is the mineral of salt, which is composed of sodium chloride (NaCl). When halite occurs as beds of solid rock, it is commonly called "rock salt." Halite is found as crystals usually white but can appear in a variety of colors due to impurities. Halite is an evaporite mineral that is very common worldwide, that is deposited in solid underground masses, and as a dissolved solution in many arid-region inland lakes and oceans. Salt deposits are mined for common table salt (NaCl), which is in great demand and has many uses, including food seasoning, road safety in the winter months to melt snow and ice, water softeners, salt licks for horses and cattle, medicinal purposes and primary ore for both sodium and chlorine. Halite is also an important component in the production of detergents, bleach, soaps, dyes, and chemicals.

Salt is obtained from two sources; *rock salt,* which is crystallized salt produced by evaporation of seawater, and *salt brine,* which is water containing a high concentration of salt (>15% NaCl). Rock salt occurs as beds of sedimentary deposits formed by evaporation and precipitation of halite from seawater in ancient marine environments. Salt deposits are found in two environments; as subsurface or surface deposits. Subsurface deposits are buried evaporite sediments that occur as ancient bedrock while surface deposits of salt occur in many inland evaporative lakes in desert regions, such as the Great Salt Lake, Utah, and Searles Lake in Mojave Desert, California, USA.

8.6.7 Talc

Talc is a hydrated magnesium silicate ($Mg_3Si_4O_{10}(OH)_2$) which is widely used not for its chemical composition but for its unique physical properties. The mineral is usually white, platy and soft (talc is the softest known mineral), with a pearly luster and soapy feel when touched (Fig. 5.6a). Other properties include being hydrophobic (water repellant) and chemically inert, and does not generate a chemical reaction when applied on the skin or ingested, except if it contains asbestos impurities. Talc is also organophilic, which means it has affinity for certain organic compounds; a property, for example, that is useful in paper making and as an adsorbent. In its natural form, talc is usually white, but can be green, grey or brownish, depending on its impurities. A rock composed primarily of talc is called steatite, while with some impurities it is called *soapstone*.

Talc is an industrial raw material in a wide range of applications, particularly as a major ingredient in cosmetic powder and other personal care products. Other applications include its use as filler in ceramics, paints, paper, plastics, rubber, fertilizer and food, and in biological wastewater treatment. Talc powder is used as a carrier for insecticides and pesticides and as a lubricant in high temperature applications. Talc deposits are formed by low grade metamorphism of pre-existing mafic and ultramafic rocks (including serpentinite), and magnesium carbonates, at low to medium temperature and pressure.

8.6.8 Zircon

Zircon is a silicate mineral with a chemical formula of $ZrSiO_4$, and is the main source of zirconium. Zircon is an industrial mineral as well as a gemstone. In its natural state, zircon forms well-shaped crystals usually white or colorless, but can show diverse colors due to traces of certain impurities. Zircon has some physical properties that include high bonding capacity, low acidity, low thermal expansion coefficient, high chemical resistance, and good recyclability. These unique properties make zircon extremely suitable for use in foundry and refractories (furnace linings and foundry bricks), and for casting applications and fabrications where its resistance to shock, abrasion and high temperature make it a valuable material. For ceramics, zircon dioxide powder is added to glazes for floor tiles and sanitary ware.

World production of zircon in 2019 was estimated as 1.6 million tons, of which the leading producers were Australia (550,000 tons), South Africa (350,000 tons), United States (100,000 tons), China (80,000 tons) and Mozambique (50,000 tons). In Nigeria, zircon occurs as an accessory mineral in the alkaline and peralkaline granites of the Jurassic Younger Granite Complex in Plateau, Kano, and the Nasarawa States, and also as xenocrysts in the Cenozoic alkali basalts in the Mambila region and Jos Plateau.

8.7 Description of Selected Industrial Rocks

Industrial rocks are rocks (indurated aggregates of minerals) that are used as raw materials for the manufacture of industrial products or for ornamental applications without any mineral separation. This definition distinguishes *industrial rocks* from *industrial minerals*. Apart from the metal mining industry, the ornamental stone industry is the oldest known mineral industry that used earth materials. The best-known industrial rocks are limestone, phosphate, coal, diatomite, carbonatite and nepheline syenite, and ornamental stones -marble, granite, serpentinite and soapstone.

8.7.1 Limestone

Limestone is a common sedimentary rock composed mainly of one mineral, calcite (or aragonite), a carbonate of calcium ($CaCO_3$). Limestones are usually white in color, but may range from buff to gray due to impurities (Fig. 8.4a). Limestones are formed in marine environments by two processes: (1) biochemical or biogenic processes; and (2) chemical or inorganic precipitation from sea water due to evaporation. Limestone is a versatile industrial rock used in a variety of industrial applications, including the following: as a major component in cement manufacture; pulverized limestone powder used as soil conditioner; production of quicklime (calcium oxide) and slaked lime (calcium hydroxide); for use as fillers in paints, paper, plastics, rubber and textiles, and for production of cosmetic products, baking soda, pharmaceuticals, toothpaste and as an additive in glass manufacture and binder for slag in blast furnaces for iron purification. Limestone is mostly extracted from surface quarries. Limestone is present in most Phanerozoic sedimentary basins around the world. In Nigeria, limestone deposits are found in most sedimentary basins and used as raw material at about a dozen cement factories.

Fig. 8.4: (a) Limestone (b) Phosphate rock

8.7.2 Phosphate Rock

Phosphate rock or phosphorite is a chemical sedimentary rock composed mainly of phosphate minerals that occur as compact or nodular masses (Fig. 8.4b). The dominant phosphate minerals are varieties of the mineral apatite: fluorapatite, ($Ca_5(PO_4)_3F$,) chlorapatite ($Ca_5(PO_4)_3Cl$), and hydroxylapatite, ($Ca_5(PO_4)_3OH$), usually admixed with other detrital materials. Variety of textures include; oolitic, nodular, pelletal, or micritic, and closely mixed with calcite, quartz, and clay minerals. The phosphates are formed by biochemical precipitation of shells of marine invertebrates containing calcium phosphate, and the bones and excrement of vertebrates such as birds. Phosphorites are deposited mostly in shallow marine or low-energy environments. Thick deposits occur as beds intercalated within limestone, shale and carbonaceous mudstone.

The phosphate content of phosphorite varies greatly, from 4% to 35% phosphorus pentoxide (P_2O_5), but typical phosphorites contain about 30% P_2O_5. The grades of phosphorite are classified as high-grade (>30% P_2O_5), medium-grade (20%–30% P_2O_5), and low-grade (15%–20% P_2O_5). The very high- grade ore blends with medium-grade ore for direct sale. The low-grade ore is beneficiated to about 28% by flotation techniques to remove impurities. Phosphate is used worldwide for agricultural fertilizer and animal feed supplements with 10% for industrial chemicals and other applications. Togo is a leading producer of phosphate, while Nigeria has a few occurrences of thin beds of phosphate in the Sokoto basin at Dange and Dahomeyan basin at Ifo, Ogun State.

8.7.3 Basalt

Basalt is a mafic volcanic rock that is used in the production of stone wool. Stone wool or mineral wool is an insulation material formed by spinning molten rock and minerals with steel slag to create a cotton-candy-like wool product (8.5a). To produce stone wool, basalt with added dolomite is melted like lava in a cupola furnace heated to over 1,500 °C through which a stream of air or steam is blown. More advanced production techniques utilize the spinning of molten rock in high-speed spinning heads to create a cotton- candy-like wool product. Pressed into rolls and sheets, stone wool creates incredibly effective insulation with sound-absorbing and fire-resistant properties and can be used for ceiling tiles and all types of sound or thermal insulation and fire protection in many applications. It is particularly suitable for industrial use given its high- temperature resistance and non-combustibility. Several countries produce mineral wool including Nigeria.

Fig. 8.5: (a) Basalt Rock Wool (b) Ornamental Marble Decor

8.7.4 Ornamental Rocks

Ornamental rocks are those rocks used for decorative purposes in construction, architectural design and monumental sculpture (8.5b)
. These include external and internal decorations, such as columns, house trimmings, flooring, countertops and monumental work. The rocks are also cut into slabs and tiles for wall and floor decorations. The ornamental stone industry is a longstanding industry that has a large potential market worldwide and has become an important industrial sector for many European countries. The most common ornamental rocks are *marble, granite, serpentine* and *soapstone.* It is pertinent to note that ornamental stones used for decorations are distinguished from "building stones" that are cut into certain dimensions as blocks or crushed and used in building construction (walling, roofing and cladding).

8.8 Description of Selected Construction Materials

Construction materials are a group of *unconsolidated* or loose earth materials and crushed or cut rocks that are used directly in building and road construction. Examples include sand, gravel, crushed rock aggregates, laterite, brick clay, building stones such as slate and sandstone, and cement materials.

8.8.1 Sand and Gravel

Sand and gravel are naturally occurring materials that are found in stream beds or terraces (8.6a). Sand is a fine- to medium-grained water-borne or wind-borne granular material made of fine-grained rock and minerals of a size less than gravel (<2mm) but coarser than silt. Their properties usually depend on the composition of parent rocks. Sand grains can be sharp ('sharp sand') or rounded ('smooth sand') depending on the angularity of the grains. Sand is usually excavated from buried or current river beds, as well as dredged from coastal areas where beach sands are available. After excavation, sands may be washed to remove undesirable material or 'fines', or screened to separate or divide into desired size groupings. Sand is a major ingredient of concrete, plaster, mortar and road pavements.

Gravel is a naturally occurring, water-borne or water-worn pieces of rock that are >2mm in size and found in old or current stream beds. They are normally rounded with smooth surfaces, and their properties may be dependent on parent rock. Gravels of varying sizes are mixed depending on the level of sorting. River gravels are usually made up of minerals or rocks that are resistant to weathering, such as quartz, quartzite and some fine-grained mafic rocks. Prior to use, gravels may be washed to remove undesirable material or may be screened to separate into desired size groupings. Most countries of the world have their own sand and gravel supplies. which are derived locally for building and construction.

Fig. 8.6: (a) Sand and Gravel

(b) Crushed Stone Aggregate

8.8.2 Crushed Stone

Crushed stone refers to rock fragments produced from hard rocks by mechanical means such as quarrying and crushing (8.6b). Geologically, they can be derived from igneous rocks such as granite or basalt; sedimentary rocks such as limestone or sandstone; or metamorphic rocks such as quartzite and gneiss. Crushed stones are produced in a quarry or mine. After being extracted from a rock face by blasting, and are processed through several stages of crushing and screening to create the final product. Crushed stone is a major ingredient in various types of concrete including cement concrete, asphalt concrete, and as ballast for railroads. The most common rocks used for crushed stone are granite, granite gneiss and marble.

8.8.3 Building Stones

Building stones are naturally occurring rocks that can be quarried and cut or shaped into blocks or slabs for use in building walls, cladding, paving, roofing or other construction activities. Natural building stone is the oldest construction material, and has been used by human beings since prehistoric times. Many ancient structures built with limestone have survived as archeological remains for thousands of years. In modern times, rocks are still used as building and structural materials where they are locally available, although more are used for decorative purposes. The most common rocks used as building stones are gneiss, diabase/dolerite, sandstone, quartzite, slate, marble, limestone and granite. Laterite is also cut into bricks for building in tropical areas of Africa and Asia (Fig. 8.7).

The main attributes of a good building stone are appearance and color, weight, resistance to fire, hardness and toughness, strength and durability. Also, a building stone should be readily available, easy to quarry and worked into desired shapes. Slate is a dark or grey, usually fine grained, foliated metamorphic rock formed by low-grade metamorphism of shale, and has been used as a building material since the ancient Roman times. Natural slate is composed mainly of fine micas or clay minerals, and the relative concentrations depend on the intensity of metamorphism that the rock has been exposed to. Some of its striking properties include its fissility (ability to be split into thin slabs), attractive dark color, water resistance, fire resistance, environmental friendliness, and durability which make slate a very good choice for building construction. Quartzite and sandstone both contain quartz which is a tough mineral to crush, and are both used as building stones. Quartzite is a non-foliated metamorphic rock that is composed of almost 100% of interlocking crystals of quartz formed by metamorphism of quartz-rich sandstone. Due to the hardness of quartz and the interlocked nature of the crystals, quartzite is a hard, tough and durable rock that is relatively difficult to break or crush.

Fig. 8.7: (a)Quartzite Blocks (b) Laterite Bricks (c) Slate roofing tiles

8.9 Gemstones

Gemstone (or gem) is a special class of industrial (non-metallic) minerals which occur as well-formed crystals that can be cut and polished for use as jewelry and other ornaments. There are over two hundred minerals recognized as gemstones. For a mineral to qualify as a gemstone, it must possess certain qualities; including beautiful color and clarity (no inclusions), and hardness and toughness to be able to resist scratching. These properties make gemstones valuable, although rarity may also be another characteristic that lends value to gemstones. Gems are so valuable in their rarity, and sufficiently durable to give lasting pleasure to their owners. Although almost all gemstones are minerals, there are a few rocks (such as lapis lazuli) and organic materials that are not minerals (such as amber, jet, and pearl) but are also used for jewelry, and are therefore often described as gemstones.

There are two main types of gemstones, based on their quality and value: *precious* and *semi-precious* stones. Precious stones are diamond, emerald, ruby and sapphire. Semi-precious stones include tourmaline, garnet, amethyst, malachite, azurite, aquamarine (beryl), rose quartz and peridot (olivine). Diamond, made of recrystallized carbon, is the most precious and alluring of all gemstones; sparkling and colorless or just faintly colored. Emerald is a green variety of the mineral beryl ($Be_3Al_2(SiO_3)_6$. Ruby is a pink to blood-red colored variety of the mineral corundum (Al_2O_3). Sapphire is a typically blue variety of the mineral corundum.

The world trade in gemstones is very substantial. In 2018, the total value of the global production of rough diamonds amounted to 17.4 billion U.S. dollars. In 2019 the value of rough diamond sales worldwide amounted to 13.9 billion U.S. dollars. It has been calculated that the world colored-gem trade is worth over$6 billion per year. Although synthetic forms of many gems now exist, they have not had a serious impact on the international gemstone market. *Gemology* is the science of gemstone that deals with both natural and artificial gem materials. It is a geoscience, and a growing branch of mineralogy. Some jewelers (and many non-jewelers) are academically trained gemologists who are qualified to identify and evaluate gemstones.

8.9.1 Precious Gemstones

Precious stones are natural gems that are visually appealing, highly valuable and attractive. This group comprises four specific stones*; diamonds, rubies, emeralds, and sapphires* (Fig. 8.8). All precious stones are translucent and are valued by the richness of their color, except for diamond, which has a higher value based on being colorless but sparkling. Their rarity, beauty, and method in which they are produced, all add to the allure of precious stones. Diamond which is made of recrystallized carbon is the most precious and alluring of all gemstones; often sparkling and colorless or faintly colored. Emerald is a green variety of the mineral beryl ($Be_3Al_2 (SiO_3)_6$. Ruby is a pink to blood- red colored variety of the mineral corundum (Al_2O_3) while sapphire is a typically blue variety of the same mineral corundum.

Fig. 8.8: Botswana Diamond Mozambique Ruby Madagascar Sapphire

8.9.2 Semi-Precious Stones

Semi-precious stones refer to all gemstones other than the four precious stones. Examples of semi-precious stones are shown include garnets, tourmalines, topaz, tanzanite and zircon (Fig. 8.9). Semi-precious stones are considered as less valuable or less appealing, although attitudes have changed regarding what gemstones are precious today. Some semi-precious stones have been prized higher than some of the so-classified precious stones. For example, the three most valuable and prized gemstones today are tanzanite, black opal and red beryl.

Fig. 8.9: Tanzanite Nigeria Topaz South Africa Zircon

8.9.3 Mineralogical Groups

Gemstones can be classified into mineral groups called "species". The most important groups in terms of variety and color are Beryl, Corundum, Diamond, Garnet, Tourmaline, Quartz and Zircon. Most of the gemstones have a Mohs hardness of about 7 and above.

Review Questions

1. Define industrial minerals and list their main characteristics. Briefly explain their importance in the national economy of developing countries.
2. Describe how the physical or chemical characteristics of the following industrial minerals allow for the commodity to be used by society: (a) Barite (b) Diamond (c) Ball Clay (d) Gypsum.
3. What are industrial rocks? How are they distinguished from construction materials? Select one industrial rock and describe its uses and origin.
4. Describe the various types of industrial clays, their industrial properties and industrial applications.
5. What are gemstones? Briefly describe their unique attributes Attempt a classification and list the seven mineral groups (species)

Chapter 9: Mineral Exploration

9.1 Introduction

Mineral exploration is simply the systematic search for mineral deposits, and is essential for ensuring that the world has a stable and sustainable supply of minerals to meet human needs now and in the near future. Mineral exploration has become a very important aspect of economic geology because it provides an organized process for locating new mineral deposits, and also serves as a link between *mineral deposits geology* and *mining* (extraction of minerals). The search for mineral deposits is not new, and has been practiced since ancient times when minerals became prominent commodities for the use of mankind. Ancient Mayan Indians in South America and Egyptians in North Africa mined copper and gold respectively, and developed methods of finding copper ores based on visual prospecting. What has changed in modern times is the use of modern technology in aiding the search for mineral deposits. Mineral deposits are the foundation of human civilization and are essential to human existence because most of the household and industrial products that we need or desire for a good quality of life are produced from metallic and non-metallic minerals.

Mineral exploration is undertaken by all types of private organizations (companies, partnerships, corporations), and public (government) agencies, and some programs are funded by international organizations, such as the United Nations and the World Bank. Since mineral resources are non-renewable, there is a continuing need to find more minerals to satisfy human consumption. However, the discovery of new minerals has become more difficult in recent times because the easy to find and near-surface minerals have already been discovered. So, for economic geologists involved in the search for mineral deposits, methodology has become more sophisticated, and the use of advanced technology has become necessary to aid in the search for new mineral discoveries.

9.1.2 Definition of Terms

The terms *mineral prospecting* and *mineral exploration* are often used synonymously but they are not exactly the same.

Mineral Prospecting

Mineral prospecting is a process of conducting a field search for minerals of economic value by examining outcrops and field exposures. It is a small-scale activity run by one or few individuals (prospectors) looking for minerals using traditional methods. Mineral prospecting may be integrated with the reconnaissance phase of a mineral exploration program. Compared to mineral prospecting, mineral exploration is a much more advanced, intensive, systematic, organized and professional form of ore search.

In many parts of the world, such as Canada, Norway, Australia, Finland or the United States, prospecting for minerals used to be carried out by "laymen" or non-professionals called "mineral prospectors". They generally don't have significant training in geology, but are usually field-oriented individuals with interest and some knowledge of minerals, who go out into the field, using visual techniques to find useful minerals. Sometimes they work with or accompany field geologists as field assistants. (The author had the opportunity and experience of working with prospectors in the search for uranium in northwest Canada). The prospectors are able to stake claims on mineral properties as practiced in countries like Canada and the United States

The traditional prospectors searched for valuable minerals by direct observation of mineralization in outcrops, panning of sediments, and soil whereas modern prospectors of today are better equipped by acquiring more knowledge about rocks and mineral identification, and adopting more technology that employs a variety of portable field equipment which can help identify economic minerals, such as the Geiger Counter for radioactive minerals, metal detectors for iron-bearing minerals etc. In developing countries, such as in Africa, there are local people in the villages who have "a good eye" for minerals, and are sometimes called "pathfinders". They assist prospective miners and geologists in locating minerals such as gold, gemstones, cassiterite in soils and alluvial gravels. A lot of artisanal miners depend on such help in finding extractable minerals. During the mid-19th century in North America, prospectors were very common, and they were the individuals who made the "gold rush" in California so monumental.

Mineral Exploration

Mineral exploration can be defined as "an organized and systematic process of searching and finding mineral deposits, and evaluating their economic potential". Another relevant definition of mineral exploration is: "the process of finding commercially viable concentrations of minerals for profitable extraction". In practice, mineral exploration often involves a complex and sequential process of information gathering and analysis, and using such information to locate and evaluate potential concentrations of mineral deposits. Mineral exploration is more robust than mineral prospecting and can be subdivided into preliminary, intermediate and advanced stages if necessary. A modern mineral exploration project handled by big corporations is usually an organized and large-scale operation with the following steps:

- Identifying what minerals and what areas will be the focus of ore search
- Finding the mineral resource
- Evaluating the ore potential for mining
- Extracting minerals at a profit

In the making of a mineable ore deposit, the second step of finding the mineral occurrence or deposit is by far the hardest part of the process. The third step of conducting preliminary and detailed evaluation of the discovery usually takes several years and a lot of work to determine the economic prospects of the mineral discovery. If the mineral occurrence is already known, then the first and second steps are skipped, and the focus is placed on the third step of detailed exploration.

9.2 Stages of Mineral Exploration

Mineral exploration is a complex set of activities that involves systematic phases similar to a "life cycle" that starts with developing an idea of what mineral to look for and ending in a mineral discovery, evaluation and mine development. The following stages have been recognized in the mineral exploration cycle (Fig. 9.1):

1. Program Planning
2. Desk Study
3. Reconnaissance Exploration and Prospecting
4. Intermediate Exploration
5. Advanced Exploration
6. Feasibility Study and Mine Planning

The activities at every stage are summarized in Table 9.1. As a rule, positive results in any stage of exploration will normally result in an advance to the next stage and an increase of the exploration effort and expenditure. Some authors have used a different scheme to describe the stages of the mineral exploration cycle as follows: (1) area selection, (2) target generation, (3) resource evaluation, and (4) reserve definition. This scheme is almost equivalent to the scheme proposed above for the mineral exploration cycle. The difference is just a matter of emphasis, and not of scope. The latter scheme is used commonly by mining engineers, and lacks some of the detailed activities important to economic geologists.

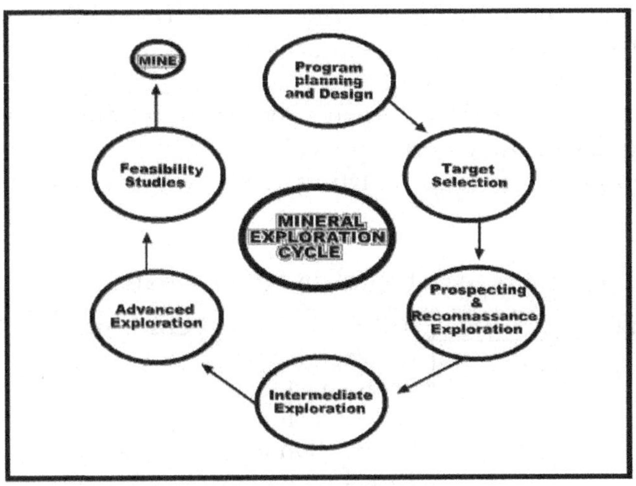

Figure 9.1: Mineral Exploration Cycle

9.2.1 Program Planning

Most mineral exploration projects typically begin with *project planning* which is undertaken by the management and professional staff, along with the gathering and review of publicly available information on various aspects of the proposed exploration project. During the initial planning effort, geological and economic issues of critical importance are decided. Geological issues include identifying what areas will be the focus of the exploration (target area), and selecting mineral(s) and ore deposit type to be targeted. Economic considerations will vary widely depending on the type of mineral deposit sought, the nature of the target area (e.g. remoteness, access, and infrastructure), market conditions and a proposed budget. Most of the time, the basic economic goal of mineral exploration projects is to find ore deposits at minimum costs and extract them at maximum profit.

The choice of what *target mineral(s)* or ore deposit type to look for may be determined by several factors, including market economics (market demand, metal prices) and a company's preferred or strategic interests that are more or less dictated by existing policy. For example, a government-funded exploration agency may decide to explore for iron ores because of the need for raw materials for a proposed iron and steel industry (e.g. Nigeria), or to locate uranium ore for nuclear energy reactors to supply power... The choice of the *target area*(s) is often left to the technical team of experienced geologists and geophysicists who are knowledgeable and capable of making such decisions, as well as choosing the appropriate exploration methods and the applicable field technologies.

Target area selection (or targeting) is the process of selecting the area(s) with a good prospect of containing mineral deposits of interest. This is a critical step in mineral exploration decision making, because it will obviously impact the ability of the exploration program to succeed and find ore deposits easily and quickly. Several factors affect the choice of target areas, most of which are *geological*, but some are *economic* (e.g. taxation, trade practices, stability), *political* (e.g. instability, leadership quality), and *social* (e.g. insecurity, unrest, civil wars). Other minor but related concerns are metal prices which may be too low or fluctuate widely, and the issue of remoteness and lack of infrastructures (e.g. power, transportation). Most of the above considerations are straightforward business-related decisions that are usually made by companies or agencies in accordance with their philosophy and policy.

Geological factors are of utmost importance in selecting target areas for exploration. They are affected by the company's approach to exploration and exploration strategies. First of all, a choice may be made in the selection of either the greenfield or brownfield approach. In *brownfield* exploration, target selection is relatively easier because the potential targets will be located in areas near or adjacent to known mineral occurrence (s)t or an already operating mine, whereas greenfield exploration requires starting in a new area selected based on geological concepts.

Table 9.1: Stages and Summary of Activities in Mineral Exploration

	Exploration Stage	Activities and Techniques
1	**Planning**	Database search Economic modelling Economic forecasting Exploration model review Target mineral/metal selection Target area selection Field Activities planning
2	**Desk study** *(Target Selection)*	Computer database research Literature search and review Remote sensing analysis Review of geology and mineral deposits reposts Review of topographic and geology maps Compilation of data and reports Land acquisition / Exploration permit
3	**Reconnaissance Exploration and Prospecting** *(Target Generation)*	Photogeology Remote sensing Regional geological mapping Aerial geophysical surveys Stream sediment geochemical survey Prospecting

4	**Intermediate Exploration** (*Target Appraisal)*	Detailed geological mapping Photogeology Rock, and Soil geochemical surveys Stream sediment surveys (small scale) Ground magnetic, electromagnetic, electrical surveys Ground radiometric survey
5	**Advanced Exploration** *(Target Testing)*	Drilling Pitting and Trenching Core logging Sampling and Assaying Geostatistics and Geological model studies Mineral resource evaluation Ore reserve definition
6	**Feasibility Study** *(Target Evaluation)*	Scoping Study Pre-Feasibility report Full Feasibility report
7	**Mining** *(Target Extraction)*	Mine design Mine development Mine production

9.2.2 Desk Study

The next stage after project planning and target area selection is to conduct a desk study. This is a process of carrying out research and acquiring information on the selected target area(s). Besides background information on geology, other relevant data on the occurrence of producing or closed mines and prospects in the area (if available) are essential. This information will normally be obtained through extensive literature review of published information, including scientific articles, reports by government agencies e.g. Geological Survey and Mines Department, and research materials (dissertations, theses, publications) on existing mines, mineral occurrences and geological characteristics of the potential target area(s). Other items such as topographic maps, geological maps, aerial photographs and Landsat imagery should be reviewed and used in developing a good knowledge of the target area. These maps and imagery may not be easily available in many developing countries where geoscientific information is inadequate.

In Nigeria, for example, topographic maps on a scale of 1:100,000 or 1:25,000 are available, but may be old and date back to the colonial era. Aerial photographic coverage and Landsat Imagery are rare or difficult to obtain. Other relevant information such as aerial geophysical (magnetic, radiometric) and geochemical maps (if available) should be examined so as to shed more light on prospective areas and help in developing a comprehensive strategy for reconnaissance exploration. At the end of the desk study and planning, the geological staff will recommend the appropriate methodologies for field exploration to the management for approval, and a time schedule and general budget are established for the project.

9.2.3 Reconnaissance Exploration and Prospecting

After the initial process of area selection and desk study is complete, the reconnaissance stage is initiated with the main purpose of evaluating the target areas of interest rapidly, and identifying potential anomalies that may become the focus of further investigation. Finding potential targets or anomalies in the reconnaissance stage often involves systematic geological investigations through field work and remote sensing analysis; aerial geophysical surveys, and if applicable, conducting wide-spaced (regional) geochemical surveys using stream sediment sampling which allows coverage of a large catchment area.

Reconnaissance exploration aims at rapid and low-cost sorting out of prospective parts of an area. It may also incorporate "mineral prospecting" which involves "putting the boots on the ground" by traversing the target area searching for valuable minerals, primarily by direct observation of outcrops, stream sediments, and soils. In areas with good rock outcrop, the first stage will be to check the geology by driving along roads or from the air if ground access is impossible. Apart from the geological hammer, portable geophysical equipment such as the Geiger Counter for radiation detectors (for uranium) and metal detectors for magnetic ores are used in spot testing for mineral showings (Fig. 9.2a &b). For heavy minerals such as gold, tantalite and corundum, panning of stream beds may apply (Fig. 9.2). Tools that are used in reconnaissance exploration include (1) remote sensing analysis, (2) reconnaissance geological surveys, (3) aerial geophysical surveys and (4) reconnaissance geochemical surveys.

Remote Sensing

Remote sensing involves gathering information about the Earth's surface by measuring electromagnetic radiation, and field signals that emanate from objects. Remote sensing allows us to collect information about an object on the Earth's surface without making physical contact with that object, and allows data acquisition before extensive field observation. Using aerial photographs, and satellite and spectral imagery (e.g. Landsat), remote sensing is *a* sophisticated but valuable tool that can be used during reconnaissance exploration to "look down" on potential geological and structural features or targets that may indicate potential mineralization.

Remote sensing images are used for reconnaissance mineral exploration in two key ways: (1) mapping and analysis of the geological and structural features (e.g. faults and fractures) that may be associated with an ore deposit. (2) recognizing unusual features such as hydrothermally altered rocks by their spectral signature. It is therefore plausible that mineralized zones exposed on the earth's surface may be detected by identified anomalous alteration zones or structural features that localize mineralization that appear as lineaments on satellite imagery. Applications of remote sensing can also provide value by reducing the risk of a project and helping prioritize which sites to explore first. The benefits of using remote sensing in exploration of remote areas where accessibility is minimal cannot be overemphasized, particularly in many parts of tropical Africa.

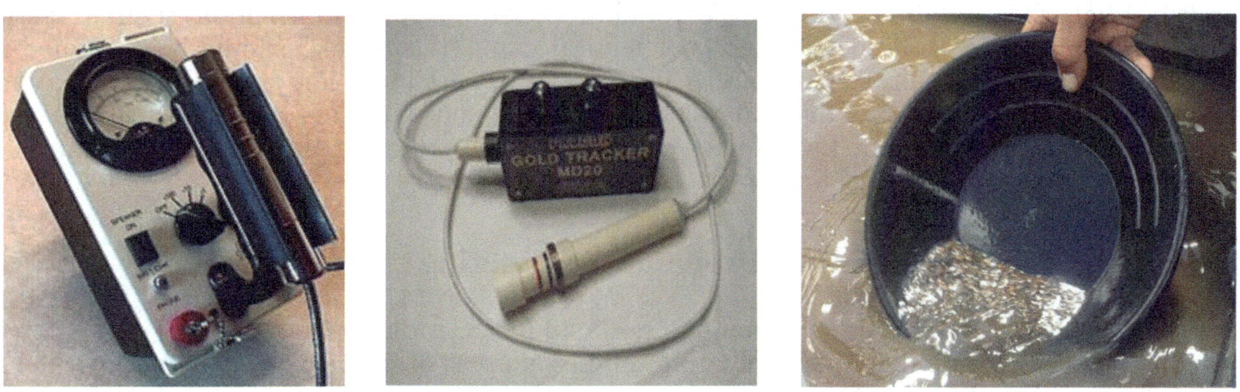

Figure 9.2: (a) Geiger Counter (b) Metal Detector (c) Panning for Gold

Reconnaissance Geological Investigations

This phase usually includes some outcrop mapping, outcrop sampling, and testing of rocks and minerals. Traversing is usually undertaken by foot, or use of field vehicles or by helicopter in areas that are not easily accessible. The preliminary field work helps to provide basic knowledge of the rocks and familiarization with the target area. It may also be carried out on a broader scale to follow up objects of interest that were identified from remote sensing analysis or aerial photographs, such as lineament traces, zones of alteration and depth of weathering. Existing geological maps provided by government geological surveys at a scale of 1:25,000 or 1:100,000 can be useful in reconnaissance geological investigation by providing knowledge on the geological setting. Rock and mineral samples collected during field visits to outcrops may be sent to the laboratory for thin sectioning and preliminary petrographic and ore microscopic studies. Systematic sampling of fresh rock samples during traverses can be analyzed as part of a preliminary lithogeochemical survey.

Airborne Geophysical Surveys

These are the most effective and widespread method of gathering reconnaissance data at this stage of exploration. Low flying winged aircraft, helicopters and drones (Fig. 9.3) are used to carry geophysical equipment to collect data on the magnetic, electromagnetic, and radiometric properties of rocks that can generate anomaly maps. Airborne geophysical survey is an obvious choice for rapidly scanning regional target areas in a cost-effective manner during initial exploration particularly in remote terrain. Several ore deposit types have unique geophysical attributes that make them detectable from airspace. The real outcome from these surveys are broad anomalies that could be attributed to several geological factors that could be filtered to identify potential exploration targets. Such targets can be followed up by ground-based surveys.

Reconnaissance Geochemical Surveys

Geochemical surveys using widely-spaced stream sediments are commonly used in reconnaissance exploration particularly in areas with adequate drainage systems and annual precipitation that allow stream flow (Fig. 9.4). Stream sediments represent the best possible composite of weathered and primary rocks over a large area. They are composed of clastic and hydromorphic materials that include clays, detrital fine-grained rock and mineral particles, organic matter, and iron and manganese coatings on clasts derived from upstream catchment areas. If mineralized areas exist within the catchment area, it is likely that some components are incorporated in the sediments. Sampling and chemical analysis of the clay/silt fraction (minus 75-80 mesh) may show secondary dispersion of ore and pathfinder elements, and geochemical anomalies following the "path" of increasing values upstream may lead to mineralization in the catchment area of drainage basins. Low sampling density such as one sample per 10 or more square kilometers are considered appropriate for reconnaissance exploration. The outcome of the reconnaissance stage may lead to a follow-up and detailed geochemical surveys. using soil and rock sampling.

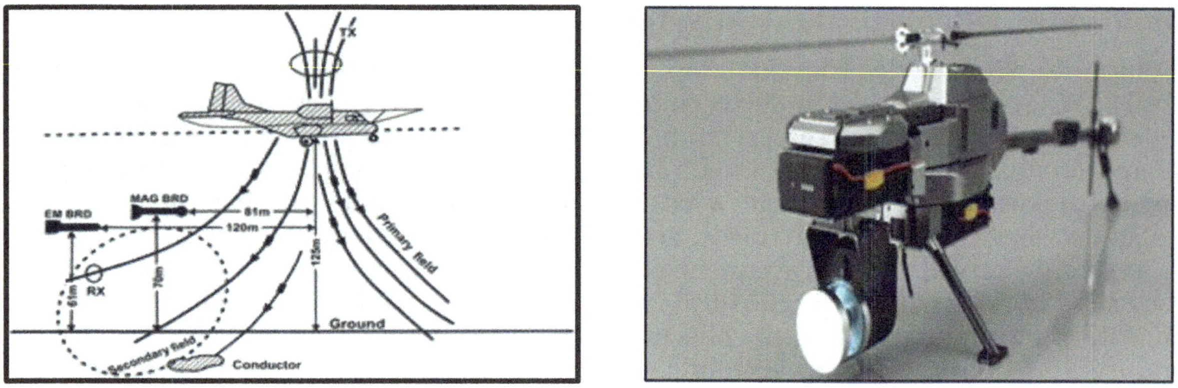

Fig. 9.3: (a) Aerial geophysical detection of orebody (b) Drone (UAV) Geophysical Surveyor

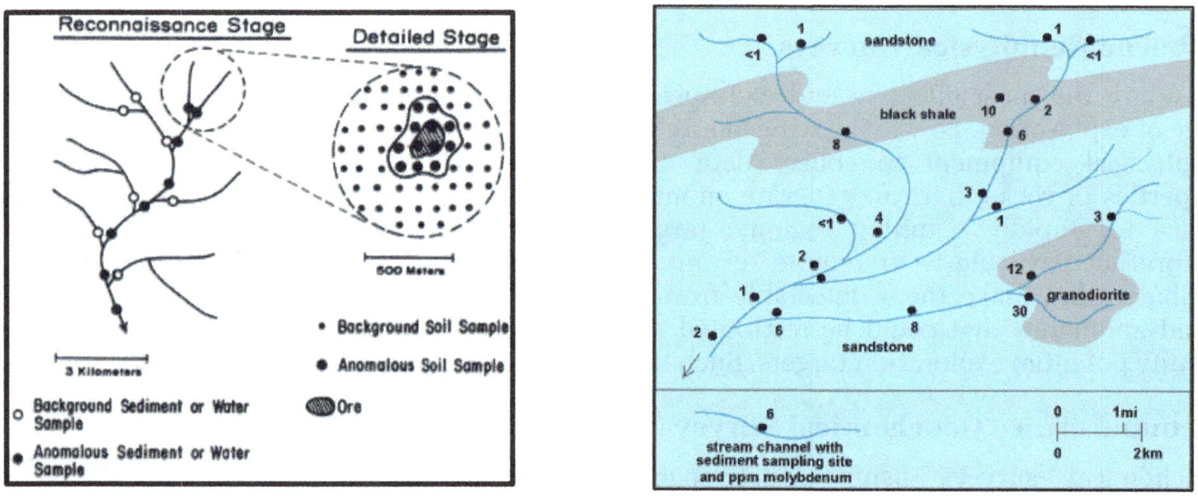

Figure 9,4: (a) Concept of reconnaissance geochemical survey (b) Zinc geochemical anomaly

9.2.4 Advanced Exploration

After the preliminary or intermediate exploration stage, a number of prospects may be identified, and the most promising of these become the focus of the more advanced or detailed exploration. This is often is restricted to relatively small areas and can be intensive and expensive, particularly where drilling is involved. The principal aim of detailed exploration is to further define and evaluate the mineral prospect(s) in detail. The objectives are to: (a) obtain information on the subsurface characteristics of the prospects, including the lithology, structural features and mineralization characteristics; (b) define the vertical and horizontal dimensions of the ore body; (c) provide samples for assay and mineralogical studies; (d) conduct mineral resource and ore reserve estimates; and determine the worth of the ore deposit.

Prior to the commencement of detailed exploration, *detailed geological mapping* of the selected areas accompanied by *pitting and trenching* may be required. Several advanced exploration techniques are used, depending on the type of deposit and its proximity to the surface, but subsurface *drilling* is foremost and accompanied by core logging, sampling, assaying, and eventually, ore reserve estimation. If results are encouraging, the project will proceed to pre-feasibility and feasibility studies, where other non-geological factors are considered to establish if a mining project will be feasible or doable based on all the data that have been accumulated. Advanced exploration g is the most important and resource-consuming phase of mineral explorations, and on the average, accounts for over 50% of exploration costs. The following activities are associated with advanced exploration. (1) Pitting and Trenching, (2) Drilling, (3) Core logging and Sampling. (4) Assaying and (5) Ore reserve estimation.

Pitting and Trenching

Pitting and/or trenching are often utilized to probe anomalies observed at the surface in a target area, if under soil cover at shallow depths, to access the potential orebody. Pitting is best applicable where the mineralized zone is flat lying, such as placer deposits or sedimentary deposit type whereas trenching is preferred for zones or rock bodies that are elongated, such as structurally-controlled mineralization (e.g. veins, dykes). Stripping the overburden and trenching or pitting are activities that can involve use of heavy machinery such as a bulldozer or backhoe to remove the overlying soils, and may require the use of explosives to blast a trench in the subsurface rock. Hand-dug pits and trenches may be used in places where power excavation equipment cannot be brought to remote locations. In very rare situations, special drifts or tunnels may be driven in order to better expose and explore a deposit, but this is a very expensive and time-consuming practice. In general, the purpose of driving such drifts is to provide drilling sites from which a large volume can be explored and a three-dimensional model of the potential ore body developed. Old shafts and drifts often provide a valuable and convenient way of exploring extensions of mineralization particularly in brownfield exploration projects.

Drilling

Drilling is the process of mechanically putting a hole in the ground that extends into the subsurface to obtain fresh and representative samples and geological information. Drilling is the most widely used and expensive tool of mineral exploration, and on the average, accounts for more than 40% of exploration costs. Drilling is used to obtain detailed information about rock types, mineral content, rock fabric, and the relationship between the host rocks and mineralization at depth. Over 95% of mineral sampling is done by drilling. Drilling helps locate and define ore bodies and provides most of the subsurface information required for the economic evaluation of a mineral prospect. In setting up a drilling program, the geologist in charge of the project will make several decisions including (a) where to drill, (b) the type of drilling required, (c) the drill hole spacing (d) the timing of drilling, and (e) the contractor to be used. The pattern of drilling used is usually based on the assumed outline, attitude and thickness of the ore body. This depends on the information obtained from geophysical data and geological extrapolations based on the deposit model which may not be accurate.

Vertical drill holes are the easiest and cheapest to drill and widely used for mineral deposits that are flat lying, disseminated or gently dipping (Fig. 9.5a). On the other hand, inclined holes are usually preferred for steeply-dipping mineralization. Drill holes are usually installed systematically and at close intervals along the strike and dip of the target deposit. A surveyed grid system is usually used to accurately determine drillhole location, according to the shape, size, and disposition of the orebody, and to facilitate mathematical calculations in ore reserve estimate (Fig. 9.5b). Although most drilling is vertical, inclined and horizontal drilling are quite common depending on the inclination of the ore body.

There are several methods of drilling but the three most common methods employed in exploration drilling are; (a) *diamond core drilling*, (b) *percussion air drilling*, and *(c) rotary drilling*, Selection of the right drilling method depends on several factors including the type of rocks, depths of the drill holes, required sample volume and quality, logistics, speed, cost, and environ-mental considerations. Diamond drilling is a preferred technique in hard rocks and for great depths. Percussion drilling is preferred for shallow depth and in unconsolidated rocks. Rotary drilling using air or mud to bring out the cuttings is suitable for exploration in sedimentary or unconsolidated sediments extending into great depth. Focus of this section will be on diamond drilling where samples can be obtained for chemical and optical analysis.

Diamond core drilling is the most widely used technique in the exploration of deposits in hard rock. It differs from other geological drilling methods in that it produces a solid core extracted from depth for examination at the surface More than any other exploration techniques, drilling provides the exploration geologist with the most concrete and accurate material upon which an economic evaluation of a mineral prospect can be made. It also provides a detailed, continuous look at the subsurface geology that may be displayed in 3D. Diamond core drilling is best used in consolidated or hard-rock formations because of better core recovery. Compared to rotary drills, diamond core drills are small in size, and are well-suited for remote exploration sites.

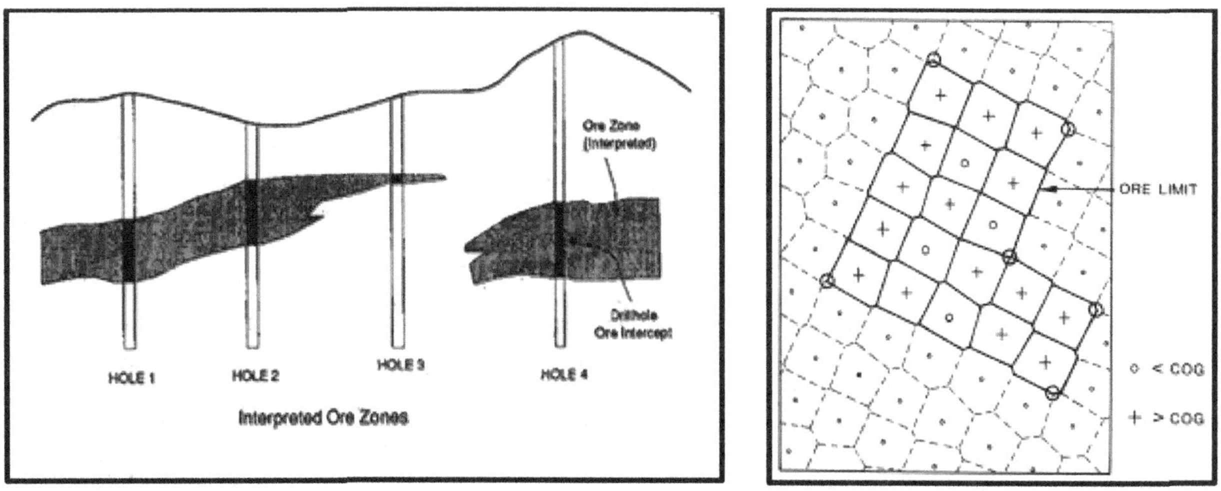

Fig. 9.5: (a) Diagram of drill holes intersecting orebodies (b) Borehole drilling pattern

The core holes in diamond drilling may be hundreds or even thousands of meters in depth, and samples can be obtained throughout the length of the holes (Fig. 9.5a). There are several core diameters but the most common (standard) diameter is about 50 mm (2 inches). The drill holes are usually arranged in a systematic pattern at fixed intervals (Fig. 9.5b). In the drilling process, a mounted drill machine is usually placed on a truck (Fig. 9.6a)), uses a diamond-tipped bit to cut a narrow kerf of rock, extracting intact a cylindrical core of rock in the center. The key technology of the diamond drill is the actual diamond bit itself. It is composed of industrial diamonds set into a soft metallic matrix. The diamonds are scattered throughout the matrix, and the action relies on the matrix to slowly wear during the drilling, to expose more diamonds (9.6b). The cylindrical drill cores extracted from the subsurface are usually preserved for future storage by being arranged in sequential order in core boxes, according to the depth from which they were obtained (Fig. 9.6c). The drill cores may be split longitudinally, and half of the split core is assayed while the other half is permanently stored for future use and re-assaying if necessary.

Figure 9.6: (a) Diamond drilling (b) Diamond Drilling bits (c) Drill Cores

Core Logging and Sampling

After diamond drilling is complete, a series of drill cores (Fig. 9.7a) set in core boxes become available for geological examination. The geological logging of other non-core materials such as rock chips from percussion drilling and auger drill samples also take place but in a different format because of the low reliability of such samples. Drill cores provide the best and most reliable samples of the mineral deposit because they represent a complete record of the rock and ores encountered in the subsurface during detailed exploration.

Core samples most commonly appear as continuous cylindrical bodies of rock (if recovery is 100%) usually of the standard 2 inches diameter but may be composed of broken interlocking pieces of rock/ore if recovery is poor. Poor core recovery often reflects zones of tectonic activity, such as faulting or shearing. Diamond drilling produces rock samples that can be examined and used to determine the mineral concentration, thickness, and depth of mineralized zones. Geochemical analyses of the drill core samples provide the basis for determining the average grade and value of the ore deposit.

Core Logging Methods: Core logging is the geological examination of drill cores involving a systematic recording and measuring of as much information as possible regarding the lithology (rock types), mineralogy, potential geological history, structure and alteration zones that occur underground through a tiny piece of cylindrical rock. Core logging is usually undertaken by the project geologist, and requires highly specialized skills. In practice, the basic information required during core logging should include the following: lithological description (e.g., color, texture, mineralogy, rock alteration, and rock name), fracture spacing and orientation, core recovery (including the location of excessive core loss, >5%), photographic documentation, description of the geological structures visible in the core, preliminary geological profile, rock properties for calculating geotechnical parameters and content and distribution of mineral and ore components, including as possible in situ assaying of ore. The description must be quantitative as much as possible and systematic (Fig. 9.7b).

The data from drill core observations are usually presented in a concise format in prose or summary forms or graphs to be utilized in the interpretation of the subsurface geology in the holes drilled or those to be drilled. The way such data are recorded influences in a significant way the amount and type of data that are observed. Using the most appropriate system for recording geological observations of drill cores is therefore very important in the understanding of the mineralization. Although a huge number of different logging forms have been devised and used by companies during exploration, there are only three basic methods for recording observations on drill core and cuttings: (a) prose logging, (b) graphical scale logging, and (c) analytical spreadsheet logging.

Prose logging uses description of observations in words for a selected interval of drill cores defined by depth limits (e.g. depth of 145 to 150 feet). *Graphical scale logging* uses standard forms that can include several mapping columns along with extra columns for recording digital data, sketches, verbal comments, etc. *Analytical spreadsheet logging* is less commonly used, but consists of a computerized spreadsheet formal using symbols to represent observed data and related explanations.

Sampling is the collection of a representative part of the mineral deposit, sometimes including the host rock. Sampling of the drill cores is usually carried out during or after the logging, and the samples obtained are sent to the laboratories for mineralogical studies and/or chemical analysis (assays). The standard practice is to split the cores in half, and one half is sent for assaying while the other half is preserved for the record and for future reference. All sampling should be carried out carefully and diligently, using scientifically established sampling practices designed and tested to ensure that the results are representative and reliable. A geoscientist should supervise the sample collection and ensure that a chain of custody of the samples is established and recorded.

Fig. 9.7: (a) Drill cores in storage (b) Cure logging

9.2.5 Assaying

Assaying is the chemical analysis or testing of a rock, mineral, or ore sample to determine its metallic contents or grade. Assays are very critical in evaluating whether a newly discovered mineral deposit has economic potential. The assay value of an ore body is the same as the ore grade or tenor. It measures the metal content as a percentage of the total weight and is expressed as a percentage, e.g., 4% iron by weight (kilograms), but for precious metals that occur in very low concentrations, different units of measurement are used, such as ounces/ton or grams per kilogram. Assays are carried out in designated laboratories using instrumental or classical methods of analysis. Considerable variability may exist in assays because of sampling errors or the difficulty of obtaining a representative sample from drill cores. Usually, the larger the core sample, the better the chances of obtaining accurate results. During sampling and analysis, quality assurance and quality control (QA/QC) guidelines should be followed to ensure reliability of results. QA/QC procedures include replicate analysis and inclusion of blanks and standards with known values in each batch of samples forwarded to a laboratory for analysis.

9.2.6 Statistical Analysis of Exploration Data

Statistical analysis of exploratory data is a major requirement in ore reserve estimation so as to understand the statistical and spatial character of variables on which the estimate depends, to detect possible errors, and also recognize any information that is useful for validating ore deposit models. Such data analysis normally includes interrelationships among variables of interest, recognition of systematic spatial variation of the variables (e.g. grade, thickness, bulk density, etc.), and identification and understanding of results that lie outside expected distributions, known as "outliers". Apart from the statistical summaries (mean, median, standard deviation, etc.), graphical summaries such as frequency distributions, probability plots, scatter plots for correlations, and various multivariate procedures such as trend analysis, multiple regression analysis (e.g. bulk density–metal relationships) and multiple variable plots (e.g. ternary diagrams) may also be computed to provide a better understanding of the exploration data.

9.2.7 Evaluation of Exploration Data by Computer Modeling

Raw data obtained from exploration needs to be carefully examined to ensure that decisions to proceed with the exploitation of a new deposit are based on objective, reliable, accurate, relevant and transparent evaluation. There are a number of steps that the project evaluation process needs to go through before major decisions are made. Computer modelling is a good tool in such evaluation.

1. Collection of geological information
2. Development of geological model
3. Integration of mineral deposit model
4. Determine mineral resource characteristics
5. Calculate ore reserves
6. Conduct Pre-feasibility study

Deposit Models

Deposit models are simulations created from the combinations of diverse geoscience information on the geological and mineral deposit characteristics of an exploration target for resource evaluation. Both geological model and ore deposit models are usually generated from available data

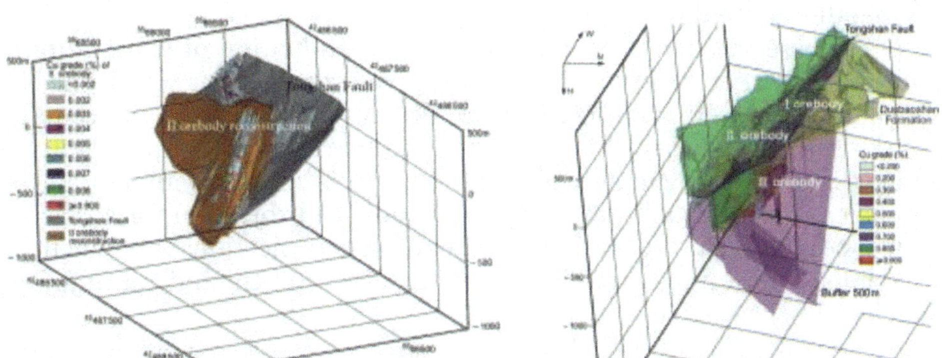

Fig. 9.8 (a) Geological block model and (b) Ore deposit model

Geological Model is a reflection of the geologists' understanding of the mineralization; its host rocks and its structural deformation (9.8a). The relevant data are collected during the long process of exploration, including results of geological mapping, rock chip samples, trench data, geophysical and geochemical data, and drill log data, and the geological model is built using 3-D modeling software. Its main objective is to define three-dimensional envelopes that reflect the geological characteristics that may impact ore localization, such as host-rock distribution, the erosion surfaces, geological contacts or faults.

Ore Deposit Model is developed to simulate the ore body boundaries and the distribution of ore grades (9.8b). The model is prepared by superimposing a digital 3-D block model or framework onto the geological model. The ore deposit may be subdivided into blocks, and the dimensions of the blocks depend largely on the size and the shape of the orebody. The model will help identify areas with the highest grades and assist in future mine planning and development.

9.2.8 Ore Reserve Estimation

Ore reserves are simply, the estimates of the quantity (tonnage) and quality (ore grade) of a mineral deposit that has been tested. After the completion of well-organized exploration activities involving drilling of several boreholes into the subsurface, and detailed examination, sampling and assay of the drill cores, ore reserve estimates can be calculated. Geological and ore deposit models are developed as described above, to provide adequate 3D representation of the information on the nature and extent of mineralization. The calculated tonnage and grade of the ore deposit may be integrated into such models to show the vertical and lateral disposition of mineable ore bodies.

Ore reserve estimation and grade determination are significant steps in the exploration and subsequent mining of a mineral resource, and have become of crucial importance due to the large investments involved in the development of a mining project. The accuracy of information gathered during exploration is very important in reserve estimates. The more accurate the exploration data, the more precision is achieved during estimation of mineral reserves. Reserve estimates are made for various reasons, including creating a basis for placing a value on a mining property for the benefit of the owners or investors, and also for tax purposes. The term "*mineral reserve*" is often used interchangeably with "*ore reserve*", but they both refer to the same thing - the value of minerals, although some authors prefer to restrict mineral reserves to non- metallic minerals and rocks (e.g. limestone, coal, feldspar, gypsum) while "ore reserve" is more applicable to deposits with metallic content (e,g. iron, copper}.

Three types of reserves are often reported during evaluation of exploration data: *proven, indicated* and *inferred. Proven* reserves are quantitative estimates for which the tonnage, grade, and orebody shape have been computed from dimensions revealed in outcrops, trenches, underground workings or drill holes. *Probable* or indicated reserves are quantitative estimates for which tonnage and grade are calculated partly from specific measurements and samples from well-spaced drill holes or production data, or partly from the projection for a reasonable distance on geological evidence. *Inferred* reserves are just qualitative projections, not usually considered in official reporting of exploration results. The proven reserves are the indication of ore that can be mined immediately.

9.3 Calculation of Ore Reserves

Ore reserve calculations are based on the results of detailed exploration activities involving drilling of several holes and chemical analyses of the samples obtained from them. Since the sample points (drill holes) are spaced at specific intervals, certain assumptions have to be made regarding the geological continuity and grade of ore between drill holes. The level of confidence in the data increases with the density of sampling. Calculations can vary from a very simple manual effort to highly complex mathematical, geostatistical and computer applications. In the simple calculation of ore reserves, the main objective is to determine the *quantity* (tonnage) and *value* (grade) of mineable ore within the deposit by using mathematical parameters such as volume and specific gravity of the ores.

There are four important parameters that are relevant to ore reserve calculation, for which reliable data must be obtained from exploratory drilling and laboratory testing. Input parameters may vary depending on the estimation method selected, but usually includes most of the following:

1. *Area* of the ore body (calculated from geometric shape used)
2. *Thickness*; (average thickness obtained from drill hole data)
3. *Average Ore Grade* (average of assay values for ore body or segments thereof);
4. *Tonnage Factor* (determined from measurements)

The simple formula for the estimate of tonnage is: (*Area* x *Thickness* x *Tonnage factor*). Ore grade is not used directly in the calculation of tonnage, but it is an important parameter in ore reserve estimates of metallic ores since the grade or content of valuable metal determines whether the deposit is viable for exploitation or not. For example, an iron deposit with 300 million tons of 75%Fe differs in economic viability from another iron deposit with 300 million tons of 32% Fe. The first deposit is minable at a profit and considered as an ore deposit, while the second deposit is not economically viable to mine under current prices and market conditions and therefore cannot be considered an ore deposit. Consequently, the tonnage of an orebody cannot be estimated without considering the issue of grade, except in non-metallic mineral deposits.

There are two broad techniques used in ore reserve estimation; *Geometric* and *Geo-statistical*. The geometric methods are also called "classical" or "conventional" methods.

9.3.1 Geometric Methods

The geometric methods use simple geometric criteria (shape, area and volume) to estimate the tonnage of ore deposit. The difference in the geometric procedures is due to the different geometric shapes or geometric function used, and the ways in which the sample data are combined. Both the tonnage and average ore grade are calculated based on the supposed geometry of the ore body, using shapes or geometric figures such as squares, rectangles, triangles, polygons etc. The widely used classical methods are (1) Tabular method (2) Rectangular method, (3) Polygon method, (4) Triangular method, and (5) Sectional method. Worked examples are presented at the end of the chapter.

Tabular Method

This is the simplest method involving a stratified or horizontal tabular body of uniform thickness, such as beds of industrial minerals and rocks (e.g. limestone) (Fig. 9.9) The tonnage is determined simply by calculating the volume (area x thickness) multiplied by the specific gravity.

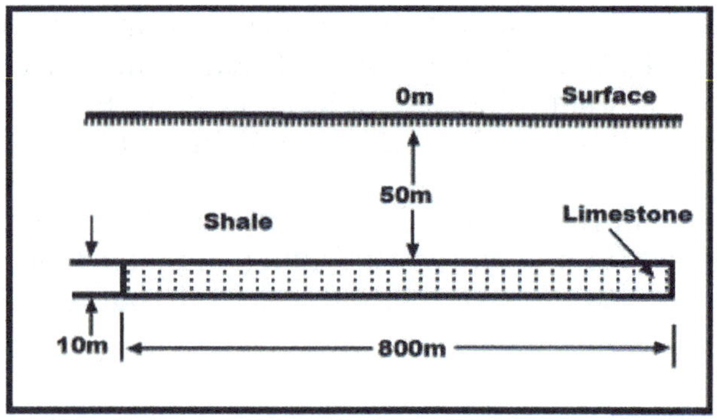

Fig. 9.9: Schematic diagram of a stratified deposit

Block and Rectangle Method

In the block or rectangle (rectangular) method, the ore body is divided into rectangles or blocks, or blocks within a large rectangle, with defined length, width and thickness, determined from drill holes located within the blocks or rectangle (Fig. 9.10). The volume is obtained by multiplying the area of each block by the average thickness of the ore in that block. Summation of the volumes of all blocks gives the total volume of the ore. This value is multiplied by the tonnage factor (specific gravity) to obtain the extent of the deposit.

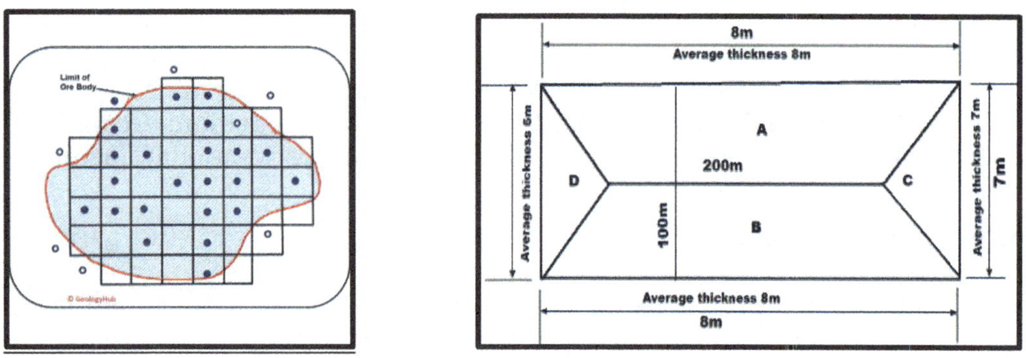

Fig. 9.10 (a) Ore body divided into blocks (b) Rectangular method

Polygon Method

The polygon method is based on a simple geometric algorithm, in which a 5-sided polygon is constructed around each hole to determine an area of influence for that hole; and then the total volume directly beneath the polygon is assigned the same values as the drill hole from which it was constructed (Fig. 9.11). The polygon method is the most widely used method, especially before the wide availability of computers. The weight for overall estimate is determined by the tonnage of the polygon that surrounds a particular sample or drill hole.

The surface area of the polygon is calculated by using the geometric function in which the *Area =1/2 x perimeter x apothem* (*1/2pa*). Apothem = a segment that joins the polygon's center to the midpoint of any side that is perpendicular to that side. The total volume is determined by multiplying the surface area with the average thickness of the ore in each polygon, and adding them up. Tonnage is then obtained by multiplying the volume by the tonnage factor in metric tons. The average grade is calculated from taking the weighted average of all grades of individual polygons that were identified.

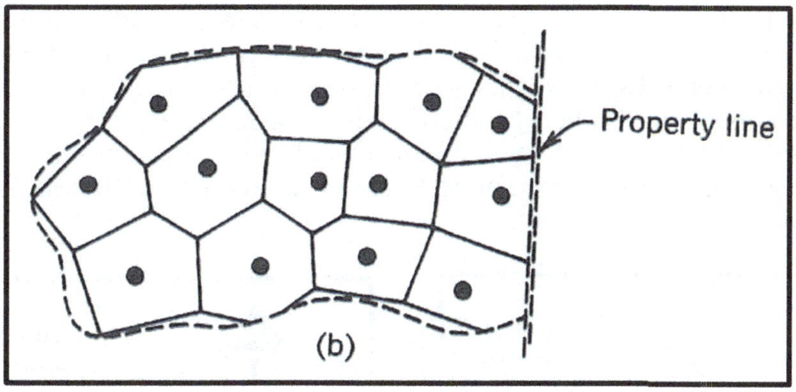

Fig.9.11: Schematic diagram of polygon method

Triangle Method

This is a modification of the polygon method but differs from it by constructing a group of triangles in such a way that each hole is situated at the corner of the triangle with the drill holes forming the three apices of the triangle (Fig.9.12). The included area of each triangle is assigned the characteristic grade not of a single hole, but of the weighted average of the three holes forming the triangle. The weighting of the three holes is based on the length of the drill holes. The advantage of this procedure is that the data in 3-point holes enter into the calculations, which then proceed the same way as in the polygon method. To determine the extent of the deposit, the area of triangles is calculated and added up. The volume and tonnage of the deposit is then determined using simple geometrical equations.

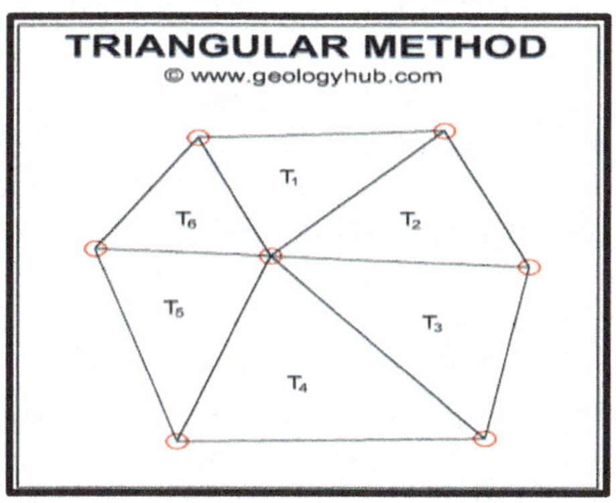

Fig. 9.12: Schematic diagram of triangle method

Cross Section Method

In the cross-section method, the ore body is divided vertically into blocks or slices of ore along regularly or evenly spaced cross-sections, and each block is defined by a drillhole intersection (Fig. 6.31) the cross sections may be vertical, horizontal, or at right angles to the dip, usually parallel to each other, and often are spaced equal distances apart. Estimates are commonly made from sections normal to the strike of the ore body. Areas of influence are calculated in terms of the area surrounding each block of ore. The limits of the blocks are set exactly halfway between the drill holes (Fig. 9.13). If the sections are quite similar in outline the volume (V), in cubic meters, is obtained by multiplying the average area (A), in square meters, by the distance between them (H) in meters, as shown by the formula: [$V = H [(A1+A2) \div 2]$] where A1 and A2 are the end areas of the block, H is the perpendicular distance between them, and V is the volume of the block. If there is a series of sections spaced equidistantly, this formula becomes V = ½ H (A1+2A2+2A3+-- An). The area of the cross-section is multiplied by the thickness of intersected ore to obtain the volume. Total volume is obtained by adding the volume of each block. Average grade is calculated from the average assays of the drillhole section. The tonnage of the deposit is thereafter calculated by the volume times the overall specific gravity.

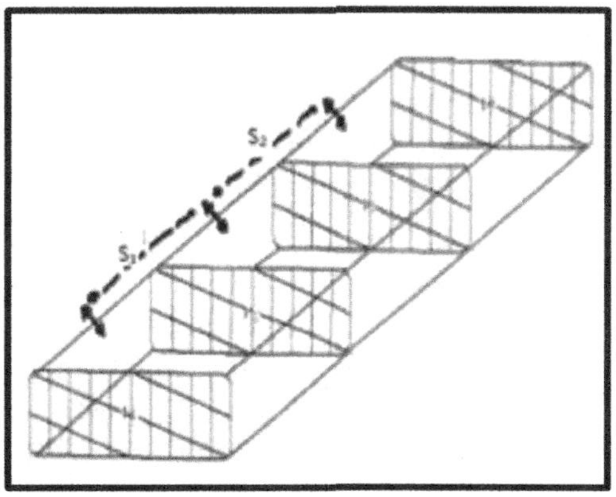

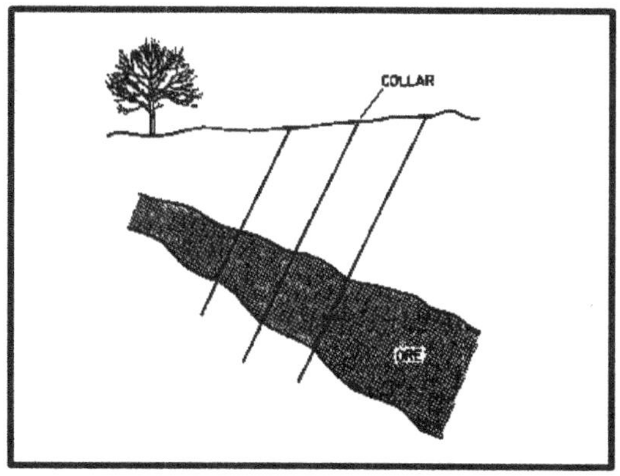

Fig. 9.13: (a) Blocks of ore in cross sections (b) Drill intersections of dipping orebody

9.3,2 Geostatistical Methods

The basic theory of geostatistics in ore reserve estimation is the application of regional variability parameters in the analysis of ore reserve data using tools such as semi-variograms and spatial interpolation procedures. In geostatistical techniques, the knowledge of the geological characteristics of the deposit is used to improve the estimate of grade, at points where no samples exist. These techniques are widely recognized as superior and more robust than classical methods. Classical statistics requires the use of a particular distribution model, e.g., the data are normally distributed or of lognormal distribution, and that the samples are independent of one another. However, in practice, there are usually insufficient samples of the ore body collected during exploration to assign a statistical distribution, and moreover, the samples are often correlated, i.e., they do not satisfy the independence requirement of classical statistics. Given that the samples, in the drill holes, are limited in number because they are expensive to acquire, often biased, and nearly always smaller in number than is desired, geostatistics has become a powerful tool for improving the quality of the estimation, using three-dimensional spatial statistics.

Geostatistical applications in ore reserve estimates follow two basic steps: *variogram* model, and *kriging*. *The variogram* model is a main tool of geostatistics. It is the first step that involves a statistical treatment of the directional interdependencies which is the quantification of the spatial variability of the ore deposit and sample grades. *Kriging* is a mathematical procedure of interpolation by assigning weights to samples so that the point estimate error is minimized. *Artificial neural network* method is one of the most recent techniques used to extend spatial distribution of sample data. The main advantage of this approach is that it requires no complicated mathematical modeling and makes no assumptions about the spatial distribution of ore grade. Details on the applications of geostatistical methods are available in the literature.

Prior to using geostatistics, the exploratory data should be statistically evaluated to determine the frequency distributions whether they are normal or lognormal (skewed), and to determine the inter-relationships (statistical correlations) among the parameters to be used in the reserve estimation. In geostatistical application, it is generally assumed that the variables in the raw data are normally distributed, and if not, a normal transformation must be applied to the data. In general, the size of the database required for ore reserve modelling and calculations is too bulky to manage manually Consequently, numerical algorithms and mathematical approaches necessitate computer applications which are currently available in many computer-aided systems and commercially available software. The accuracy and speed of computers enable evaluation of various scenarios within reasonably short times.
Geostatistical reserve estimation requires inputs that are provided in the following steps;

- Database of mineralogical characteristics, assays and geologic controls of mineralization and any other features of the deposit.
- Computation of variograms (graphical correlation of mineral characteristic) for each geologic zone.
- Division of orebody into matrices of blocks for kriging (calculation of block grade related to adjacent samples); blocks are then classed as measured, indicated and inferred.
- Estimate of tonnage and grade of each block at given cut-off-grade.
- Printing of recoverable grade distribution plans by level or bench in the proposed mine.

Comparison of estimates of tonnage and grade by hand-computed and geostatistical methods differ by less than 10% which may not be statistically significant. Indications suggest that the results obtained by classical methods are generally lower in tonnage but higher in ore grade than results from geostatistical methods.

9.4 Feasibility Studies

Feasibility studies represent the final stage of a mineral exploration project, following the completion of ore reserve estimates and the development of the orebody model. As the term "feasibility" suggests, it is an attempt to answer the question of whether a mining project will be feasible or non-doable, based on all the data that have been accumulated. It is simply a technical and economic evaluation of the possibility of developing a mine. There are three types of feasibility studies in mining evaluation:

1. Scoping Study
2. Preliminary Feasibility Study or Pre-feasibility
3. Detailed Feasibility or Full Feasibility

The *scoping study* is a summary appraisal undertaken early in the final resource assessment. The *pre-feasibility study* is a preliminary evaluation of all the parameters relevant to the decision-making process of moving a mining project forward to the next level of mine planning and design. A *full feasibility study* provides a lot of detailed information on mine design, mine infrastructures, financial aspects of raising funds for capital and operational expenses, market estimates and environmental and risk management.

9.4.1 Scoping Study

Scoping study is an initial appraisal carried out early in the process of the final resource project evaluation. It normally involves a preliminary mine development plan that forms the basis for determining whether to terminate a mineral exploration program, or determine how much further pre-development efforts are warranted. Depending on the size of the project, a scoping study can be undertaken by a single individual with expertise. A scoping study may not be accurate enough to provide a meaningful assessment of the economic viability of a project, but can determine key project risks, and help establish the probable type of mining operation, assess the potential for the delineation of resources capable of supporting a mine, and in the case of coal and industrial minerals, confirm likely markets.

9.4.2 Preliminary Feasibility Study (Pre-Feasibility)

Pre-feasibility study follows a scoping study with the intent to determine whether a mineral resource is likely to support a viable mining project. It is more detailed than a scoping study and is the key intermediate step in the assessment of a mining project. Pre-feasibility study also contains preliminary mine planning and engineering evaluations of potential mining project. The report is usually sufficiently accurate to allow a comparison of alternative project configurations and include preliminary capital and operating cost estimates. Pre-feasibility study should include, but not limited to the following:

- assessing reserves and saleable product from the delineated resources
- prepare a generic mine design
- propose non-detailed, stages in life of the mine, including planning and production scheduling, assessing the mining methods, treatment routes and identifying cut-off factors, recoveries, dilution and losses in both mining and treatment
- outlining probable plant, infrastructure, services and other facilities
- producing a summary development structure and timetable
- determining capital and operating costs
- evaluating the specification and marketability of the commodity

9.4.3 Final (Detailed) Feasibility Study

Final feasibility study is to determine whether to proceed with a mining project into the detailed engineering and construction stage. It is significantly more expensive than the preceding studies, largely because of the associated engineering work. Feasibility study deliverables will typically include, but not limited to:

- establishing the proven and probable reserves
- assessing project alternatives and selecting the desired development route
- proving the technical viability of the mine and extraction methods
- defining the mine's plant, equipment and infrastructure requirements and capacities
- establishing resource consent and other legal/governmental conditions and requirements for approvals to proceed
- defining the management control and timing of the project development and construction

- describing the commissioning of the project, recruiting and training of the management team and crew and hand over at the conclusion of construction
- estimating the development, capital and operating costs of the mine over the economic life
- identifying the market for the commodity
- assessing project risk areas, economic sensitivity to those areas and action that may be used if the risk becomes a problem
- completing economic assessments of the selected project configurations
- developing financial models
- setting a framework for the implementation of the capital investment in the mine development phase.

9.5 Worked Examples of Ore Reserve Estimates

1. EXAMPLE TABULAR METHOD

The Arimogija Limestone deposit in the Ose LGA of Ondo State, Nigeria was explored by drilling of several boreholes to delineate the extent of the deposit. Several samples were also analyzed to determine the CaCO3 content. The outline of the deposit was determined with the following dimensions: Length is 8000 meters and width is 1300 meters. Thicknesses of the limestone in four drill holes averaged 10 meters. Specific gravity of the limestone is 2.5.
Mineral Reserves (not Ore Reserves) will be calculated as Tonnage x Specific Gravity. *Volume* = 8000 x 1300 x 10 = 104,000,000 m3 x 2.5 = **260, 000, 000 metric tons or (2.6 x10^8 mT).**

2. EXAMPLE RECTANGLE METHOD

An underground mine has an orebody marked out as a rectangular block divided into trapezoids and triangles with systematic assays of the ore block at 10m intervals along horizontal drifts and vertical raises (see Fig. 9.10b). The sampling and assays were used in computing average thickness and weighted average grade, for each side of the rectangle in accordance with the length, area, or volume it represents. The data obtained is summarized in Table 9.2.

Table: 9.2: Summary data for ore reserve calculation

	Length (m)	Width (m)	Thickness (m)	Average Grade
Area A	300 + 200/2	50	6	2% Cu
(Trapezoid)	(= 250)	50	8	2% Cu
Area B	300 = 200/2	50	7	3% Cu
(Trapezoid)	(= 250)	50	6	3% Cu
Area C	100/2			
(Triangle)	(= 50)			
Area D	100/2			
(Triangle)	(= 50)			
Specific Gravity 3.0				

Tonnage in each block = length x width x thickness x specific gravity
Block A: Tonnage = 250 x 50 x 6 x 3 = 225,000 tons
Block B: Tonnage = 250 x 50 x 8 x 3 = 300,000 tons

Block C: Tonnage: = 50 x 50 x 7 x 3 = 52,500 tons
Block D: Tonnage = 50 x 50 x 6 x 3 = 45,000 tons
Total Tonnage = 225,000 +300,000+52,500+45,000 = **622,500 tons**
Weighted Average Grade = 2.23% Cu (Σ Tonnage*Avg. Grade/Tonnage)
Ore Reserve Estimate = 622,500 tons of 2.23% Cu
Copper Ore Tonnage = 622,500 x 2.23% Cu = 13,881 tons of Copper Ore

3. EXAMPLE POLYGON METHOD

A copper deposit in the middle Benue Trough, Nigeria was explored with 12 boreholes (Fig. 9.14) drilled in a grid pattern to determine the extent and grade of the deposit. The polygon method was used in the reserve estimate of the western part of the orebody in 5 polygons (D1-D5) recommended for development. Relevant information summarized in Table 9.3. The specific gravity of the copper ore is 2.7:

Table 9.3: Summary data for calculation using the polygon method

Polygon	Area (m2)	Thickness (m)	Volume (m3)	Av. Ore Grade (% Cu)
D1	500	8	4000	7,5
D2	500	10	5000	7.5
D3	500	12	6000	8.5
D4	500	6	3000	4.5
D5	500	9	4500	6.0
TOTAL	**2000**	**9 (Average)**	**22,500**	**6.8% (Average)**

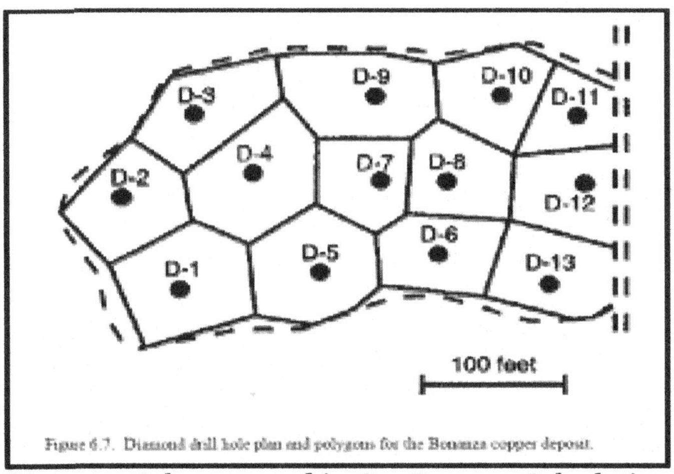

Fig. 9.14: Polygons used in Ore Reserve Calculation

As summarized in Table 9.3, the total volume of the five polygons is 22,500 m3.

Tonnage: = 22,500 x 2.7 = 60.750 metric tons of 6.8% Cu
Ore Reserve Estimate + 60,750 tons of 6.8% Cu
Copper Ore Tonnage = 60,750 x 6.8% = 4,131 metric tons of Copper ore

4. EXAMPLE TRIANGLE METHOD

The Akiri copper deposit was explored in which 10 holes were drilled (D1-D10) as shown in Fig. 9.12. The mineralized area was divided into 10 triangles T1 to T10 in which a drill hole is located at the corner of each triangle (Fig. 9.15). Calculate the ore reserves and average grade. Example of average grade calculation for triangle T1 is shown in Table 9,4 and for the other triangles in Table 9.5.

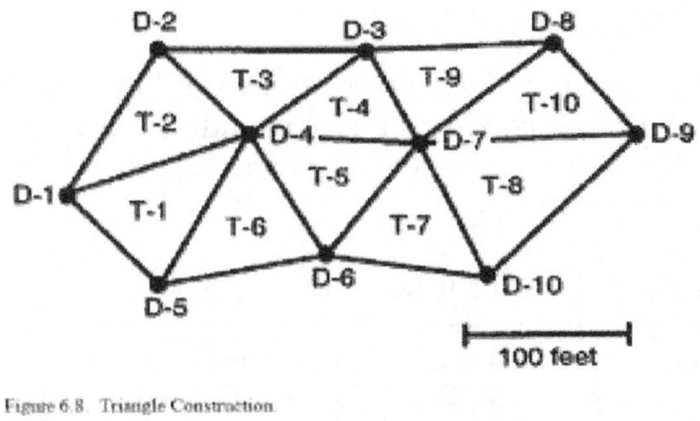

Fig. 9.15: Triangles used in Ore Reserve Calculation

Table 9.4: Tonnage Calculation for Triangle T1

Drill Hole	Thickness (m1 m)	Grade (% Cu)	Grade * Thickness
D-1	50	0.83	46.5
D-4	100	1.05	105.0
D-5	75	0.72	54.0
Total	225		205.5

From Table 9.4, Weighted Average Grade for T1 is ∑Grade*Thickness/∑Thickness =: 205.5/223 = 0.92

Table 9.5: Ore reserves parameters for the Akiri Copper Deposit

Triangle	Drill Holes	Tons	Average Grade (% Cu)	Tons * Grade
T-1	D-1, D-4, D-5	26,400	0.91	24,042
T-2	D-1, D-2, D-4	26,400	0.94	24,818
T-3	D-2, D-3, D-4	27,500	0.91	20,475
T-4	D-3, D-4, D-7	22,260	1.23	27,380
T-5	D-4, D-6, D-7	19,550	1.15	21,332
T-6	D-4, D-5, D-6	27,260	0.79	21,535
T-7	D-4, D-7, D-10	26,240	1.07	28,076
T-8	D-7, D-9, D-10	44,500	1.15	46,575
T-9	D-3, D-7, D-8	24,418	1.20	29,307
T-10	D-7, D-8, D-9	28,917	1.19	34,411
		263,445		277,927

Specific Gravity = 3.5

Given this data, the tonnage and grade calculation for triangle Tl would be as follows:

Area of triangle T1 = 4,400 square feet
Avg. Grade = 0.92
Average Thickness = 225/3 =75.1 meters
Average Grade of Deposit from Table 6.5: = 277,027/263,445 = 1.05% Cu
Volume = Area x Thickness = 4400 x 75 = 330,000 m3
Tonnage = Volume x Specific Gravity= 330,000 x 3.5 = 1,155,000 metric tons
Tonnage of Ore = 1,155,000 x 1.05% = 12,127 tons of Copper

5. EXAMPLE CROSS SECTION METHOD

A tabular dipping copper vein as shown in Figure 9.16 is sampled by two drill holes per section and one crosscut per section. The cross-section blocks B and C are spaced 100 ft apart. The tonnage factor for the copper vein is 9.5 cubic feet per ton. The following data is provided in Table 9.6.

Table 9.6: Summary data for sections

Section Block	Area (Sq Ft)	Average Grade (% Cu)	% Cu/ sq ft
B C	3260 4896	1.81 1.75	5901 8563
Total	8156		14,463

From Table 6.6, Average Block Grade =$\sum$ Area x Average Grade/ Average Grade
= 14463/8156 = 1.77% Cu
Formula: for Volume: $V = (At + A^{o})/2 * H$, (where H is spacing between blocks)
Volume of Ore Blocks = (Area B +Area C)/2 x Spacing (Spacing = 100)
Volume = = 3260 + 4895 x 100 /2 = 407,800 cu. ft. Tonnage Factor =1/ 9.5 (English Imperial)
Tonnage of Blocks = Volume /9.5 = 407,800 /9.5 - 42,926 tons of 1.77% Cu

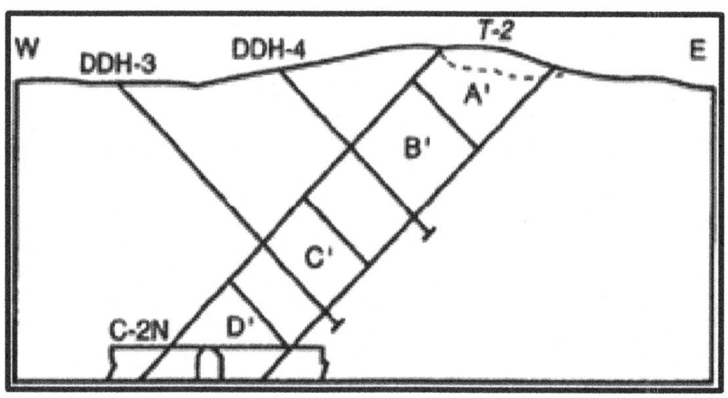

Fig. 9.16: Diagram of cross section blocks and reserve estimate

Review Questions

1. Define and differentiate mineral exploration and mineral prospecting. Briefly describe the stages involved in mineral exploration.
2. Discuss the main types of drilling in mineral exploration. Explain what mineral deposit types are suitable for each method.
3. What is sampling and assaying? Highlight the types of core logging methods.
4. Describe geological and ore deposit modeling and expatiate their importance in resource evaluation.
5. Briefly describe the various methods of mineral reserve estimation. Calculate the tonnage of a coal bed with the following dimensions: Length 10,000m; Width 500m and Thickness 10m. Specific gravity of coal is 2.2

Chapter 10: Mining of Minerals

10.1 Introduction

Mining is the process of extracting valuable minerals or other geological materials from the Earth's surface or subsurface. This is the culmination of a process that probably started with mineral prospecting, discovery, exploration, evaluation, feasibility study and mine planning and development. Mining of minerals is a human activity that started long time ago in ancient times, and has served as a source of essential metal and mineral products used by human society to improve the quality of life. The nature of mining creates a potential negative impact on the environment, both during the mining operations and for years after the mine is closed.

Mining of minerals is usually undertaken by a group of people employed in mines and known as *miners*. They operate the machinery and equipment to dig, load and transport ore and waste rock from surface and underground mines. *Mining engineers* play a professional and supervisory role in designing mines and managing operations to ensure that mineral extraction is carried out safely and efficiently.

Mining geologists assist mining engineers by providing advice on the quantity, quality, distribution, behavior and other properties of geological materials in mines, including host rocks and ores. Mining or mine geologists also provide supporting documents and vital services such as geological and geotechnical mapping, ore sampling, assays, structural analysis, guide on blasting points, environmental monitoring and ore grade control. There is a group of professionals known as *geological engineers* (not engineering geologists) who are a 'mixed breed' of geologists and mining engineers. They are better trained in geology than mining engineers, and in engineering courses than geologists

10.1.1 Brief History of Mining

Archaeological records indicate that mining commenced in prehistoric times as a result of the need of ancient people to use stones as tools for hunting. Primary stone, particularly *flint*, a hard siliceous rock, was the first earth material mined because of its conchoidal fracturing which allowed it to be broken into sharp-edged pieces that were useful as scrapers, knives, and arrowheads. However, the earliest known underground mine for a specific mineral in the world was in the Ngwenya mountains of today's Swaziland, Southern Africa where ancient (Paleolithic) humans mined hematite to make the red pigment called ochre used in burial ceremonies and as body coloring, and was dated as 43,000 years ago by radiocarbon method. Between 40,000 and 20,000 years ago, there was some evidence of coal mining in South Africa.

Mining did not become a significant industry until more advanced civilizations developed around 10,000 to 7,000 years ago when metals of copper, gold and mercury were found and valued. The first societies to use metals on a large scale emerged in Mesopotamia around 6500 years ago and Egypt around 6000 years ago. Gold was one of the first to be utilized, being mined from streambeds of sand and gravel (placer). Copper which occurs in native form was probably the second metal discovered and used. Silver was also found in a pure state and at one time was valued more highly than gold. The Egyptians and Sumerians smelted gold and silver from the gold mines of Nubia (now southern Egypt and Sudan) which were among the largest and most extensive of any in Ancient Egypt. As a result, these metals began to have a value that was transferable between people and between cultures.

Approximately 5,500 years ago in the history of mining, came the discovery of tin. Tin, mixed with copper, made bronze, the first alloy harder than its constituent metals. Metal working advanced over the next few thousand years, as did the use of other mined materials. With the development of the Mediterranean civilization, mining became one of the most critical industries in the world. Athens in Greece grew wealthy, due to the extractions from its silver mines until the fall of Athens when Sparta wrested away control of the silver mines. However, it was during the Roman Empire that mining boomed, and the Romans introduced large-scale mining operations that included surface, underground and hydraulic mining. The growth of civilization expanded rapidly, and required more and more money to finance its operations, as well as more metals to create military equipment for its ongoing wars and build infrastructure.

In medieval Europe, mining as an industry underwent dramatic changes mainly focused on the extraction of copper and iron. Other precious metals were also used, mainly for making coins. Initially, many metals were obtained through open-pit mining, and ore was primarily extracted from shallow depths, rather than through deep mine shafts. Around the 14th century, the growing use of weapons, armor, and horseshoes greatly increased the demand for iron. One of the most complete early records of mining methods in Europe is by the German scholar Georgius Agricola considered as the "father of economic geology" in his book titled *De re metallica* (1556). He described detailed methods of driving shafts and tunnels. Soft ore and rock were laboriously mined with a pick and harder ore with a pick and hammer, wedges, or heat (fire setting).

Great progress in mining was made when the secret of black powder explosives reached the West, probably from China in the late Middle Ages. This was replaced as an explosive in the mid-19th century with dynamite. During the rise of Europe in the 17th to 18th centuries, governments gave miners extensive rights to take land and profit from the minerals they found, and pay a portion of the revenue from mining as payments for granting the mining rights. Civilization progressed requiring an increasing amount of mined material to manufacture the necessities of everyday life and fund the economies of nations.

The influx of the mineral wealth from the New World comprising the United States, Canada, South America, Africa and Australia financed the Renaissance., period of political and economic rebirth in Europe. Eventually, the creation and expansion of countries in the New World, along with the development of the Industrial Age, resulted in the growth of the mining industry we have today. Coal and other metals are still essential to our day-to-day lives, and their extraction and production remains one of the most important industries in the world. A truly international pursuit, modern mining techniques are providing emerging economies with an opportunity to join the world market in a major way. A focus of modern mining practices is the use of the most efficient, safe, and environmentally friendly methods possible.

10.1.2 Mining Geologist - Roles and Responsibilities

The terms "mining geologist" and "mine geologist" are used interchangeably in this book, although in my opinion, there may be a subtle difference: a *mine geologist* is defined as a geologist employed in mines, while a *mining geologist* is a geologist with formal training, preferably with a post-graduate degree in mining geology or mining engineering. The role of a mine geologist may vary from mine to mine, but the main roles are to advice on the geological aspects of development and production processes in mines and quarries, and provide geological support to the mining personnel to ensure that economic minerals are extracted in a way that maximizes profit and minimizes problems as much as possible. The responsibilities of a mine geologist are primarily focused on the acquisition and processing of geological information including geological mapping and logging, in support of mining operations, and the supervision of drilling, sampling and ore grade control processes. The mine geologist must work closely with the mining engineers and miners during production, as well as the drill and blast engineers and mine surveyors to determine the most appropriate way to blast the rock and ore exposed on the pit floors and mine faces. These responsibilities do not end when the extractions begin, but must continue with monitoring of mine surfaces and the petrographic study of ores and materials as they are extracted, so as to assess their quality, characteristics and continuity.

During mine operations, mine geologists must continue to reevaluate the orebody, revise ore reserve estimates, and expand their knowledge and understanding of the relationships between the geology and ore formation, so as to be able to locate new mineral resources or extension of existing ore bodies. They should always be ready to provide valuable advice on short and long-term production plans, minimize dilution as well as advice on geotechnical stability of mine rocks, groundwater seepage and environmental contamination. Mine geologists must have computer skills to store and retrieve data, and use software for geological, ore reserve and ore grade modeling.

Fig. 10.1 (a) Mine geologists in underground opening (b) Mapping of open pit wall

10.1.3 Life Cycle of a Mine

A *mine* is an opening excavated in the ground either at the surface or subsurface for the purpose of extracting valuable minerals and other earth materials, usually from an ore body. The life cycle of a mine begins with discovery and ends with closure, and consists of five stages (Fig. 10.2):

1) Mineral Exploration
2) Mine Planning

3) Mine Development
4) Mine Production
5) Mine Reclamation

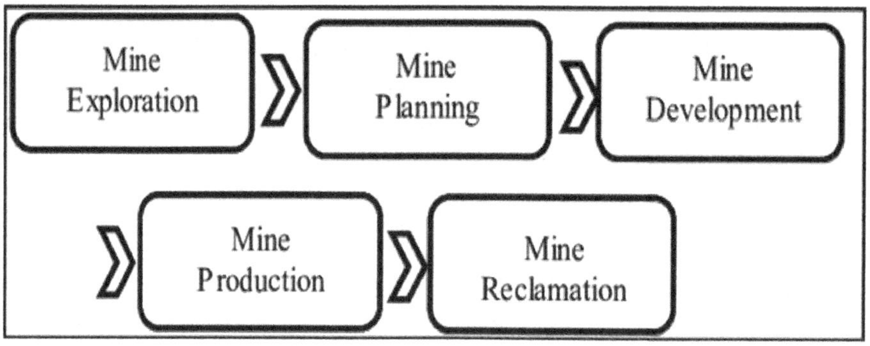

Fig. 10.2: Life cycle of a mine

10.2 Mining Methods

Mining methods are those techniques used in extracting minerals from the ground. They can be grouped into two main types: *surface* methods and *underground* (subsurface) methods. The selection of an appropriate mining method is a major decision that is often made during the feasibility study phase of mineral exploration. It depends on several factors including geological, technical, economic, environmental, and cultural. They are summarized as follows.

- *Geological* – depth to ore, thickness of overburden, shape of orebody, structural features and other geological factors such as topography and drainage.
- *Technical* – nature of equipment, size and efficiency; safety of use, and production rates.
- *Economic* –type and value of mineral resource, cost of investment, and economic viability.
- *Environmental* – toxicity of waste, proximity to water resources, impact on surrounding landscape and environment.
- *Cultural* – workforce availability, social and cultural responsibility.

10.3 Surface Mining Methods

Surface mining is the process of extracting ore minerals (and overburden) from near the surface of the Earth. It is the most preferred mining method, and is responsible for the bulk (estimated 70%) of annual mineral production worldwide. Surface mining is usually a two-step operation: (1) removal of overburden (stripping), and (2) mining of the orebody. Surface mining is typically used for surface or near-surface and usually less valuable deposits. Surface mining can be classified into two groups on the basis of the method of extraction; (a) mechanical extraction, and (b) aqueous extraction. Also, the various methods differ in terms of the mine geometries, the techniques used, and the minerals produced. *Mechanical extraction* methods employ mechanical processes in a dry environment to recover minerals from the ground. There are four main mechanical extraction methods: (1) quarry mining; (2) open-pit mining, (3) strip (opencast) mining; and (4) auger mining. *Aqueous extraction* methods use the access to water or an aqueous mixture to extract minerals during mining and processing. They recover the valuable minerals by jetting, slurring, dissolving or melting. The aqueous methods can be further subdivided into placer mining (panning, sluicing, hydraulicking, dredging) and in situ (solution) mining and leaching (Table 10.1).

Table 10.1 Classification of Surface Mining Methods

	Mechanical Extraction (Hard Rock Deposits)	**Aqueous Extraction (Placer & Salt Deposits)**
1	Quarry Mining	Panning and Sluicing
2	Open-pit Mining	Hydraulic Mining
3	Strip Mining	Dredge Mining
4	**************	Solution Mining
5	****************	In Situ Leaching

In surface mining, the first step in the production cycle is providing access to the mineralized zones by removal of the overburden (stripping) which may include rock, soil, and ecosystem that lie above the surface. *Stripping* generally involves the use of mechanized devices for excavation and transportation of the waste material to a dump site outside the area of operation. To determine how much waste material is removed relative to the ore mined, a ratio has been formulated, known as the *stripping ratio* which is defined as the tonnage or volume of overburden or waste rock to be removed, divided by the tonnage of ore to be mined from the pit. For example, a stripping ratio of 2:1 means that 2 tons of waste rock needs to be mined to mine 1 ton of ore. Generally, an open pit mine has a higher stripping ratio than an underground mine. The stripping ratio for metal mines is usually between 1:1 and 3:1 but can exceed 10:1 in mines with high-grade ore of considerable value. Stripping ratios greater than 20:1 occur in some coal mines, but are only workable if the operation is profitable. One important factor that influences the stripping ratio is the shape of the ore body which may vary considerably during the life of the mine.

10.3.1 Quarrying

Quarrying is the extraction of minerals and rocks from an open excavation or pit by digging, excavating, cutting or blasting. Quarrying is the simplest form of surface mining that is used commonly for extracting industrial minerals and rocks, and construction materials such as sand and gravel. Quarrying is used when the rock is close to the surface of the land or crops out as an inselberg or on a hillside. and building materials, such as dimension stone, construction aggregate, riprap, sand, and gravel. Based on the nature of materials extracted, quarrying is of two types. The first type is the extraction of large ornamental stone blocks of specific color, size, shape, and quality - an operation requiring special and expensive production procedures (Fig. 10.3a). For ornamental stone quarries, the common rocks extracted are marbles and granites. The second type of quarries applies to the recovery of sand, gravel, and crushed stone for the production of aggregates for road base, cement, and concrete (10.3b). Quarries of the second type are usually located close to urban areas where the products are in demand.

In the geological investigations for new quarry sites, focus should be placed on locating outcrops of the appropriate rock types that are exposed above ground and not weathered. Other factors for consideration include thickness of overburden, water table conditions, intensity of joints and fractures, nature of exposure, size of reserve, accessibility and absence of forest cover. Water flows into the quarry from the surface and from the ground may require dewatering. Many quarries are dug in hillsides so that water can drain better. For sand and gravel quarries, riverbeds or terraces provide the best locations. Laboratory tests for determining suitability as aggregates include, abrasion test, crushing test and angularity test, for quarries close to urban areas, special measures are taken to minimize environmental impacts such as noise from drilling and blasting, and dust control from crushing and screening.

Figure 10.3: (a) Quarry for marble dimension stone (b) Quarry for crushed stone

10.3.2 Open-Pit Mining

Open-pit mining is a surface mining technique of extracting economic minerals and rocks from the earth's surface by their removal from an open pit. Other terms that are used to describe this technique are *open-cut mining* or *opencast mining*. Open-pit mines are used when deposits of commercially useful minerals or rock are found near the surface. Typical ore deposits mined in open pits if exposed at the surface are vein-type, pipe-shaped (e.g., diamond in kimberlite pipes), stratiform deposits, residual deposits and widely-disseminated deposits (e.g. porphyry type Cu-Mo or Au deposits). Copper, gold and iron deposits account for over 90% by volume of open-pit excavations worldwide.

Open-pit mines are excavated in horizontal layers called *benches* (Fig. 10.4a). Each bench is a horizontal surface of defined height and width. The deposit is mined in vertical steps by adding more benches, although mining can be carried out on a number of benches at the same time, and a pit of any depth can be excavated. The top of each bench is equivalent to a working level, and access to different levels is gained through a system of ramps. The width of a ramp depends on the equipment being used, but typical widths are from 20 to 40 meters.

Mining on a new level is begun by extending a ramp downward and a new cut is then progressively widened to form the new pit bottom. As excavation proceeds, successive benches which are arranged in circular form around the nine become smaller in diameter than the last one above, and narrower to maintain safety and stability inside the mine. With time, a cone-shaped geometry is produced in which the bench area becomes progressively smaller as the pit deepens with the walls formed by cascading benches that may sometimes be steeply dipping if the rocks are very competent and stable.

The size of open pits varies considerably, and includes some of the largest mines in the world. The world's largest open pit diamond mine near Mirny in Siberia is 540 m (1600 ft) deep, 120 km (75 miles) across and yields approx. 2 million carats in diamonds per year (Fig. 10.4b). The area above the mine is a no-fly zone since it creates its own downdrafts (pulldown) which can endanger low flying aircraft. The Bingham Canyon Mine, near Salt

Lake City, is the world's deepest man-made open pit excavation for porphyry copper. The mine is 2.75 miles (4,5km) across and 0.75 mile (1,2km) deep.

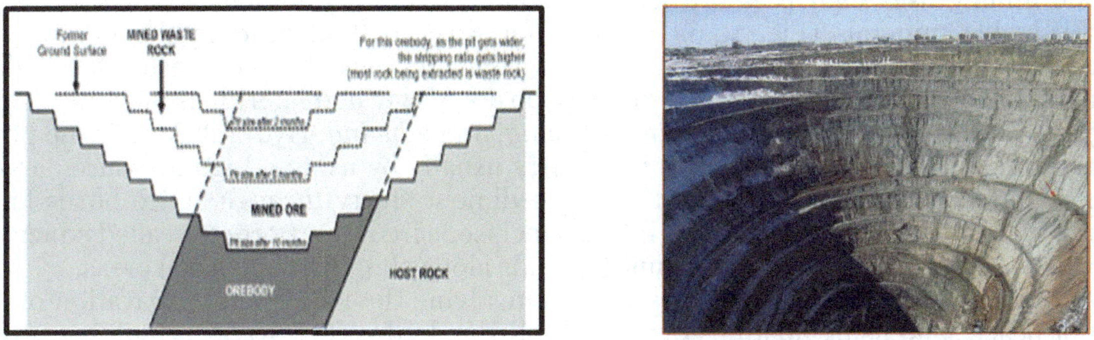

Figure 10.4: (a) Development of benches in open pit (b) Deepest open pit mine in the world

10.3.3 Strip Mining

Strip mining or open-cast mining is a form of surface mining that involves stripping the surface away from the mineral that's being excavated (usually coal), and the overburden is backfilled into the mined area. The technique is suitable in the mining of shallow, large, tabular or stratified residual and sedimentary deposits such as bauxite, phosphate (Fig. 10.5a), evaporites, tar sands, manganese, but is most commonly in the mining of coal and lignite deposits (Fig. 10.5b).

The mining process often involves first removing a long strip of overlying soil and rock (overburden) to expose the underlying mineralization or seam which is excavated en masse. The overburden is dumped directly into previously mined areas rather than outside the pit, compared to open-pit mining. The process of excavation, loading and transportation to the conveyor belt is handled by a single machine which are draglines or continuous excavators, notably the "bucket wheel excavator". These are some of the largest and most powerful machines for mining soft rock and unconsolidated deposits, and can move as much as 12,000 cubic meters of earth per hour. (Fig. 10.5a). *Contour stripping* is used almost exclusively in mountainous regions particularly in mining coal seams that crop out from the sides of hills or mountains. (Fig. 10.5b)

Figure 10.5: (a) Strip mining of phosphate, Togo (b) Contour strip mining of coal, Appalachia

10.3.4 Hydraulic Mining

Hydraulic mining uses high pressure jets of water to dislodge sediment or rock material that contain valuable minerals (Fig. 10.6a). It is a low-cost method used in the mining of large quantities of unconsolidated material such as alluvial, eluvial and beach placer deposits, particularly of gold and tin, as well as residual deposits of clay and lateritic iron oxides. Hydraulic mining involves three steps: *hydraulicking, sluicing*, and *educing*. Hydraulicking is the process of breaking up material and suspending it in a slurry usually by using a high pressure, large water cannon, followed by a process of diverting the sediment slurry through sluice boxes known as sluicing. Finally, the slurry is channeled into an enclosed circuit by a process called educing. Also, the method is applied in the mining of mineral sands along shoreline or coastal areas.

Dredging is a form of aqueous extraction involving the underwater excavation of a placer deposit or beach sand using floating equipment called a dredge (Fig. 10.6b). Dredging is the most common method of large-scale extraction of placer deposits, particularly around islands where the water is shallow. There are two types of dredges; *mechanical* or *hydraulic*, depending on the mode in which the material is transported to the surface. The mechanical dredges are excavators that utilize a mechanical method using steel "buckets" on a circular, continuous bucket line at the front end of the dredge (Fig. 10.6b). Hydraulic dredges use suction to suck placer material through a giant hose from the seabed or bottom of a river using water as the transport medium. The power used in suctioning comes from a jet-power intake. After material is suctioned up to the surface, it usually gets processed through a floating sluice at the surface. Any waste material from the box is then usually discharged over the side of the dredge.

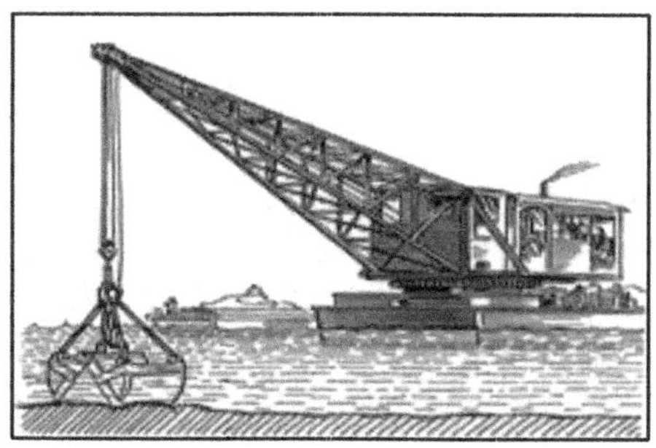

Figure 10.6: (a) Hydraulic Mining (b) Dredging of Placer Gold in Sediments

10.3.5 Solution Mining

Solution mining is the process by which useful minerals are dissolved into solution and removed from the ground. It is most applicable to water-soluble deposits of sodium and potassium evaporite salts and sulfates. Mining is performed by pumping aqueous solutions into underground wells that dissolve the whole solid minerals into a liquid. This liquid containing the minerals is then pumped out of the ground and the mineral can then be recovered by various techniques. The method has also been applied to the extraction of uranium ores hosted in porous sandstone, where the uranium salts are flushed into solution. Solution mining is more economical, safer and more environmentally friendly than conventional underground mining.

In-situ leaching is also a form of solution mining in which solutions such as acids are passed through heaps of loose rock and consolidated rocks below the surface to selectively leach useful metals. Copper and gold are extracted in this way from heaps of waste rock (heap leaching).

10.4 Underground Mining

Underground or subsurface mining is the extraction of ores from deposits that occur beneath the surface that cannot be mined by open-pit methods due to technical or economic reasons. Such deposits are usually buried and can only be reached by digging tunnels or shafts into the Earth's crust (Fig.10.7a). The main objective of underground mining is to extract ore from the subsurface as safely and economically as possible while minimizing the rock waste that is generated. Underground mining is most applicable to ore deposits that occur deep below the surface, usually where the overburden is thick or the mineral occurs as veins in hard rock. For minerals that occur deep below the surface - where the overburden is thick or the mineral occurs as veins in hard rock - underground mining methods extract the valued material most efficiently and cost effectively. Sometimes, ore deposits that exist at or near the surface may extend to greater deposits which may necessitate a transition from surface mining to underground mining.

Fig. 10.7: (a) Underground mine tunnel (b) Underground mining in South Africa

Underground mining methods are complex and varied because some were developed to satisfy the needs of different shapes of orebodies and the geotechnical characteristics of host rocks (Fig. 10.7b). The classification of underground mining methods is complex because of the unique ways in which the orebodies occur and the mining method had to be adapted to fit prevalent conditions. However, some parameters have been developed, both geological and mining related, to help classify underground mining methods including the type of access shafts used, the extraction method, morphology of the ore body, and the extent of support required. Mining in hard and soft rocks may require different techniques, due to the fact that hard rocks are more stable under unconfined mining conditions than soft rocks. Also, underground mining can involve *bulk* or *selective* methods to reflect whether the ore is recovered with or without dilution respectively. Bulk methods are used to extract low-grade and large deposits while selective methods which are more extensive are used in mining high grade metal deposits and narrow veins.

Using the extent of ground support, underground mining methods are classified by mining engineers into three groups: *unsupported, artificially supported*, and *naturally supported (caving)* (Fig. 10.8). Geologists prefer to classify mining methods using the shape and type of orebody but such classification is rather complicated and will not be used here.

Unsupported methods are essentially naturally or self-supported and require no major artificial system of support to carry superincumbent loads, instead relying on the natural competence of the walls of the openings and pillars. Examples of these methods are room-and-pillar mining, shrinkage stoping and sublevel stoping. *Supported* methods require substantial amounts of artificial support to maintain stability of its mine openings, as well as the need for systematic ground control throughout the mine. Examples of mining methods are cut-and-fill stoping, sublevel stoping and square set stoping. Naturally supported methods include longwall mining, block caving and sublevel caving.

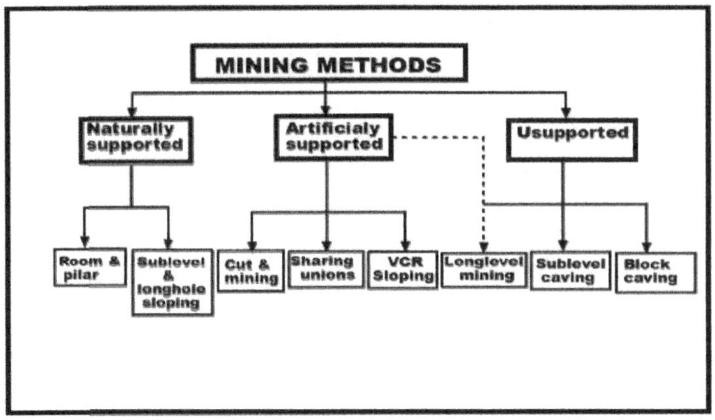

Figure 10.8: Classification of Underground Mining Methods

10.4.1 Underground Mine Infrastructure

Underground mines are complex structures with varied infrastructure and a variety of underground workings. Geologists must be conversant with the distinctive terminology that is used to describe these features. The infrastructure of underground mines is more complex than that of the open-pit mines, and can be classified into three groups: (a) access openings; (b) haulage openings and (c) development or production openings (see Fig. 10.9).

Shaft is a vertical or sub-vertical excavation sunk adjacent to an ore body in which elevators are used to transport people and ore in and out of the mine.

Skip is the portion of the shaft used for ore transport.

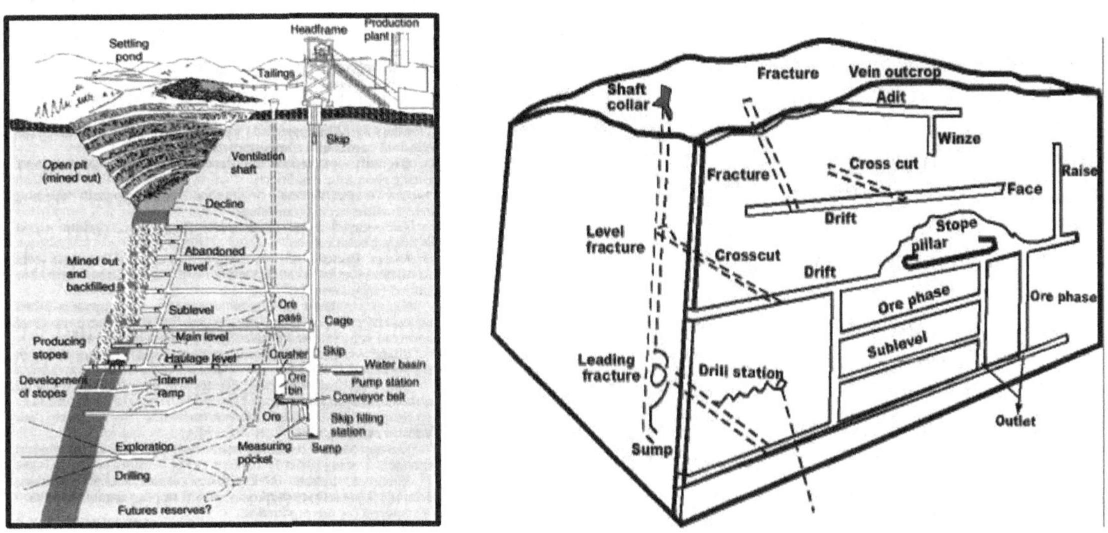

Fig. 10.9: Major Infrastructure of an Underground Mine

Ramp or Decline is an inclined tunnel sunk at a low slope angle (<20° dip) with a portal to the surface.
Adit is a horizontal or inclined tunnel excavated into the side of a hill or mountain to provide access to an ore body. The surface entrance to an adit is called a *portal.*
Manway is underground opening that is intended for personnel access and communication
Level is a system of horizontal underground workings connected to the shaft. A level forms the basis for excavation of the ore above or below.
Sublevel is a system of horizontal underground workings located between the main development level for more effective drill and blast control.
Raise: Underground opening driven upward from one level to a higher level or to the surface; a raise may be either vertical or inclined.
Winze is a vertical or inclined underground opening driven downward from one level to another level or from the surface to a level
Drift is a horizontal or nearly horizontal underground opening developed on the underground levels along the strike of the ore body.
Crosscut is a horizontal underground tunnel intersecting the ore body across its strike. The crosscuts are usually developed to connect the drives with the area in which stoping occurs
Chute is a loading arrangement that utilizes gravity to move material from a higher level to a lower level.
Cone is a funnel-shaped excavation located at the top of a raise used to collect rock from the area above.
Stope is the underground excavation made by removing ore from surrounding rock.
Face is the surface where the mining work is advancing.
Back: Roof or overhead surface of an underground excavation
Slot is a vertical or inclined ore section excavated to open up for further stoping.
Pillar is a block of ore or barren rock left intact in the mined out stope or between two stopes to act as a means of support.
Footwall is wall or rock under the deposit
Hanging wall is wall or rock lying above the deposit
Draw Point is a place from which ore is extracted from the stope and loaded for haulage by trucks or conveyor belt.
Ore Pass is a steeply dipping underground opening used for passing ore from one level to another level.
Wall Rock is the wall in which an ore deposit is enclosed.
Waste is the barren rock or rock of too low a grade to be mined economically.

10.4.2 Room and Pillar Method

Room and pillar method is the most commonly used in subsurface mining, and is best suited for mining of flat, bedded or slightly inclined (stratiform) deposits. In this system a series of parallel drifts are driven, with connections made between these drifts at regular intervals. Ore is removed from these drifts to form rooms while leaving pillars in place to provide 'natural' support to the hanging wall or roof of the room but in some mines, after mining has reached the deposit's boundary, some or all of the pillars may be removed. The goal is to recover the maximum amount of ore, and leave the smallest possible pillars. The roof must remain intact, and rock bolts are often installed to reinforce rock strata (Fig. 10.10).

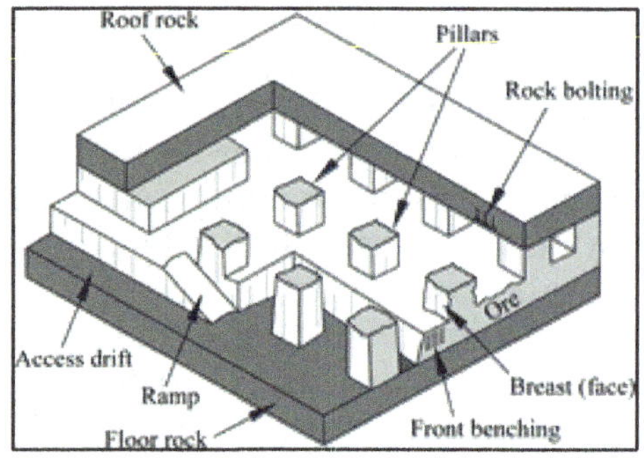

Fig. 10.10: (a) Room and pillar mine, Kansas (b) Block diagram of room and pillar method

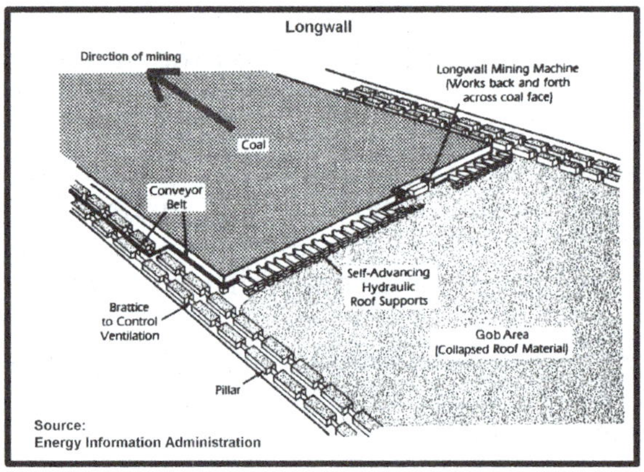

Figure 10.11: (a) Longwall mining of coal (b) Longwall grinding of coal seam

10.4.3 Longwall Mining

Longwall mining is used mainly to extract ore from thin-bedded deposits of uniform thickness and large horizontal extent (Fig.10.11a). Typical deposits of interest are sedimentary coal seams, potash layers, or fossil placers in pebble conglomerates, such as those in South African gold mines. Longwall mining can occur in both hard and soft rock as the working area along the mining face can be artificially supported where the hanging wall tends to collapse. Longwall mining is a bulk mining technique in which the orebody is divided into panels or blocks, and a cutting machine which moves back and forth is used in grinding a long straight ore surface, followed by mechanized removal and transportation by chain conveyor that carries the mineral to the haulage drift

10.4.4 Cut and Fill Mining

Cut and fill is a mining method in which an opened stope from which ore has been removed is back-filled with waste rock (Fig. 10.12a). This technique can be adapted to many different ore body shapes and ground conditions. Together with room-and-pillar mining, it is the most flexible of underground methods. Cut and fill is a highly selective method that is considered most suitable for steeply dipping high grade deposits found in weak host rock. In this method, mining begins at the bottom of the ore body and advances upwards. During the extraction process, the back of the excavated stope is temporarily supported with rock bolts before it is finally backfilled to form the floor of the next level of development and excavation support. Mining progression is achieved between stopes by the construction of raises driven upward through the ore body.

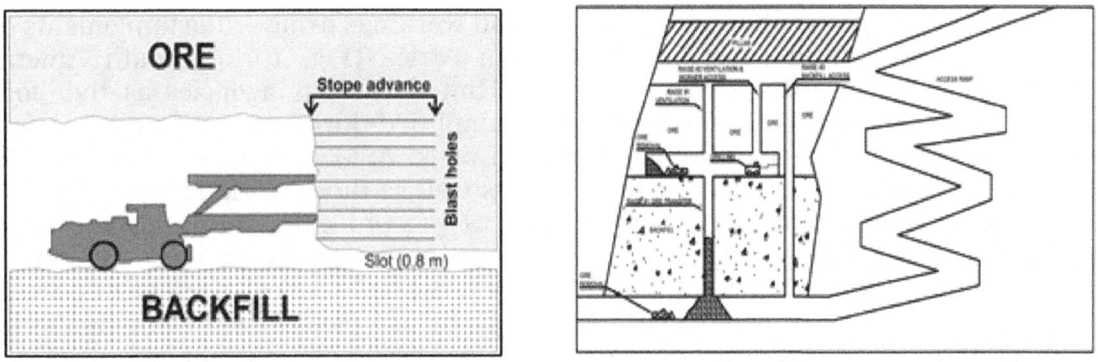

Figure 10.12: (a) Cut and Fill mining (b) Cut and Fill stoping

10.4.5 Block Caving and Sublevel Caving

Block caving is a mining method in which gravity is combined with internal rock stresses to fracture and break the rock mass into pieces that can be handled by miners (Fig. 10.13a). The ore body is divided into "blocks" or large sections of several thousand square meters, and the "caving" of the rock mass is induced by undercutting the block by blasting which destroys its ability to support the overlying rock. Consequently, gravity forces act to fracture the block while continued pressure breaks the rock into smaller pieces.

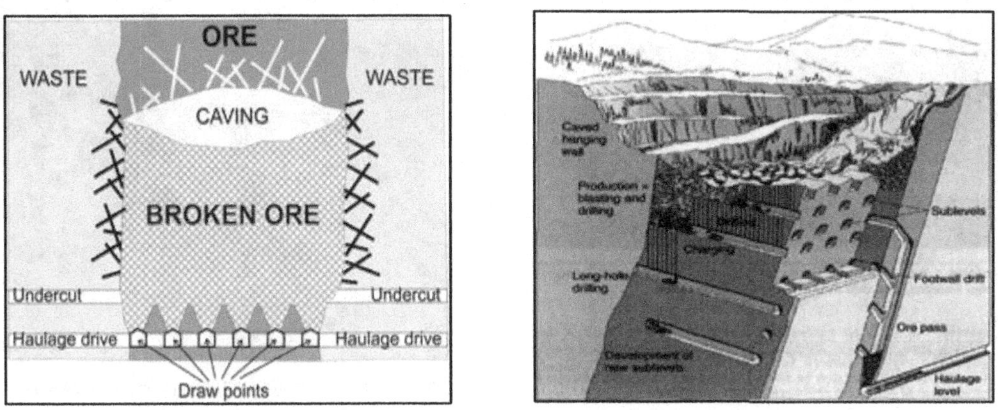

Figure 10.13 (a) Block Caving (b) Sublevel Caving

Sublevel caving is a mining method in which ore is extracted from sublevels spaced at regular intervals throughout the entire ore body (10.13b). Mining usually starts at the top of the orebody and develops downwards. A series of long blast holes are drilled in ring patterns and blasted from each sublevel in such a way to generate controlled caving when the roof fractures and collapses following the cave. The broken ore is mucked out after each blast. This mining method is suitable for extraction of ore from massive, steeply-dipping orebodies with considerable strike length, and usually has a high amount of dilution and low recoveries.

10.5 Artisanal and Small-Scale Mining (ASM)

Artisanal and small-scale mining (ASM) is a crude form of mining involving the extraction of minerals from hand-dug pits and holes in the ground by poor and often illiterate miners. This mining method involves both surface and underground workings using crude implements or "low" technology. It is common in parts of sub-Saharan Africa (Fig. 10.14), South America and Southeast Asia. Officially, ASM is defined by the United Nations agencies as the "formal or informal mining operations with predominantly simplified forms of exploration, extraction, processing tools, such as diggers, pixes, shovels and pans. ASM can include "men and women, youth and children working on an individual basis as well as those working in family groups, in partnerships, or as members of cooperatives or other types of legal associations and enterprises involving hundreds or even thousands of miners." In some countries, a distinction is made between *artisanal* and *small-scale* mining, whereby 'artisanal' typically refers to pure manual mining while 'small-scale' may have fixed installations or use mechanized equipment. However, the diversity of ASM operations is vast and generalizations are easily contradicted.

ASM is an important socio-economic sector for the rural poor in many developing nations, many of whom have few other options for supporting their families. It is widespread in approximately 90 countries in the global south from developing countries in Africa, Asia, Oceania, and Central and South America with an estimated 40.5 million men, women and children directly engaged in ASM. Twenty percent (20%) of the global gold supply is produced by the ASM sector, as well as 80% of the global gemstone and 20% of global diamond supply, and 25% of global tin production. More than 150 million people - workers and their families - depend directly or indirectly on ASM for their livelihood. 70 - 80% of small-scale miners are informal, and approximately 30% are women, although this ranges in certain countries and communities from 5% to 80%.

Fig. 10.14: (a) Artisanal mining in Congo (b) Artisanal diamond mining in South Africa

The ASM mining sub-sector operates under hazardous, labor-intensive, highly disorganized and illegal conditions which have resulted in low productivity, lack of safety (from bandits), exposure to hazardous mining conditions (pits and holes), exposure to poisonous metals (lead) and chemicals (mercury), poor health care and lack of job security (employment may be seasonal). Artisanal mining is prevalent in Africa because of the availability of a large labor force of unemployed or partially employed people that can use crude implements to extract minerals. The type of deposits amenable to artisanal mining are near surface deposits of alluvial or residual origin excavated from river beds and weathered regolith, and sometimes in weathered bedrock.

10.6 Environmental Impacts of Mining

Mining is the human activity that, perhaps, has the greatest negative impact on the environment, and human health and safety. Mining usually involves excavation of the land and rocks, both on the surface and underground. However, underground mining is a less environmentally harmful method for mineral extraction than surface mining because it occurs below the surface with limited exposure to surface water and atmosphere. The ores and other minerals that are exploited are composed of elements (metals and non-metals) that could become toxic to the environment if they ultimately accumulate in the soil or leach into surface water and groundwater that are used by humans for drinking and other uses.

Waste rock and tailings generated by mining are sometimes disposed of in a manner that is inappropriate and dangerous to human health and safety. Mine waste can pollute air and drinking water, harm wildlife and habitat, and permanently scar natural landscapes. Modern mines as well as abandoned mines are responsible for significant environmental damage throughout the mining countries in the world, particularly in China, Canada, Australia, western United States and South Africa. The impacts of mining on the environment are described in the following areas of concern:

- Land disturbances
- Safety hazards
- Waste rock disposal
- Soil contamination
- Surface water pollution
- Groundwater pollution
- Air pollution
- Ecosystem damage

10.6.1 Land Disturbance

The extraction of minerals, particularly in surface mining, usually involves the digging of open pits, and other mine workings. The excavated or dug-out materials and wastes are dumped in heaps over the surrounding lands, sometimes in a haphazard manner, thereby creating widespread disturbance of the landscape. In addition to the poor visual and aesthetic effects, mining activities have exposed the lands to degradation and erosion. In some places, deep gullies have developed from loose mine waste and abandoned pits that continue to accelerate and cut deeper into the surrounding areas thereby threatening nearby human habitat and ecosystem (Fig. 10.15a). In coal mining areas, the land disturbance can be very profound (Fig. 10.15b)

Strip mining of coal severely alters the landscape while mountaintop removal obviously destroys the natural landscape which accelerates erosion and makes reclamation near impossible. Underground mining causes less damage, and generally produces relatively small waste rock disposal areas ranging from a few to tens of acres in size. Such areas are usually located near the entrance of the underground workings, not far from the shafts and adits.

Fig. 10.15: (a) Disturbed mine landscape (b) Coal mining landscape in Appalachia, USA

10.6.2 Abandoned Mines Safety Hazard

Abandoned mines are common in the mining landscape of many countries. Many of these structures contain dilapidated frames, open shafts, and water-filled pits (Fig. 10.16a). Abandoned mines and uncontrolled mining both at the surface or underground, creates issues of public safety. Pits, holes and shafts, if uncovered after mine closure, can become death traps. Innocent people and children have been killed by falling into abandoned mine pits and underground holes while playing around or hunting. Illegal miners are known to have died when they fall into pits while digging them. Some also lose their lives while searching for mineral stones or fragments inside pits and the pits suddenly collapse, closing in on them. Another safety hazard around some underground coal mines are sinkholes or collapsed ground due to subsidence above underground coal mines. Because an unexpected collapse can occur without warning, such areas create safety concerns and usually can be identified and avoided (Fig. 10.16b).

Figure 10.16: (a) Abandoned mine (b) Sinkhole in coal country

10.6.3 Mine Waste Disposal

Mining generates a lot of waste materials, more so in surface mines that are of no economic value and need to be disposed appropriately. As the quantity of waste rock in open pit mines is commonly two to three times the amount of ore produced, tremendous volumes of waste rock are removed from the pits. These wastes include soils, rock, tailings and mineral matter usually piled up near the mine site (Fig. 10.17a). Also, waste rock and tailings usually contain metals and/or sulfide minerals in concentrations considered too low to be profitably extracted, but may be at levels that become toxic to the environment and human health. Underground mines also yield mine waste in smaller quantities that are piled near the mine access ways. Moreover, waste piles from processing, such as tailings impoundments, leach piles, and slag piles (products of smelting) can be very large, and further contribute to exposed toxic wastes. Mine waste storages are sometimes contained in dangerous structures such as embankments that could give way and lead to accidental deaths and damage to the environment and local infrastructure.

10.6.4 Soil Contamination

Soils surrounding metal mining sites are likely to be contaminated with heavy metals such as Pb, Zn, Cu, Cd, As, Ni and Se. These contaminants in soils may be products of direct transfer of mine waste and tailings or from deposits of particulates that settle from mine dust that may be airborne. Wind, precipitation, surface water and dust actions are agents that can further transport or transfer these elements to great distances far away from the mining areas. Heavy metals such as Pb (lead), Cd (cadmium), As (arsenic) and Se (selenium) and Cu (copper) are toxic if ingested or inhaled or absorbed through the skin over a long period of time (Fig. 10.17b). Such human interactions can have long term effects on health and cause diseases such as cancer, or lead to birth defects in unborn children. The area surrounding mines can also be contaminated by the various chemicals that are used in the processing of minerals, such as mercury that is used in gold extraction.

Fig.10.17: (a) Mine waste dumps (b) Copper contaminated soil

10.6.5 Surface Water Pollution

Surface water in streams and rivers near mining sites are not only disturbed but could also be contaminated by discharge of mine waste through drainage and precipitation. Sometimes the river courses may be silted or diverted by mine waste. *Acid mine drainage* which is the outflow of acidic water from mining areas, including coal fields, is the primary source of water pollution from mining. This corrosive water with low pH is derived from the leaching of ore material from the mine (mine water), waste rock or tailings (Fig. 10.18a). The yellowish-brown color of the acidic water is caused by the oxidation of pyrite (FeS_2), a common but valueless sulfide mineral, as well as other sulfides of arsenic, lead and mercury associated with metallic ores. When such sulfides are exposed to water and air, sulfuric acid is created, which can seep from mines and waste rock piles into streams, rivers, and groundwater.

10.6.6 Groundwater Pollution

Groundwater is a precious natural that is used for drinking and industrial purposes, and a source of recharge to surface water. Groundwater may be polluted by underground mining activity and its quality may be affected by the handling, storage, and disposal of mining wastes (Fig. 10.18b). Other sources of groundwater pollution include the wastewater discharge; and the storage and handling of chemicals and fuel within mines. Acid runoff from the mines is one of the most toxic sources of dissolved heavy metals such as copper, lead, mercury that may enter into the groundwater system in shallow aquifers. The rate and degree by which acid-mine drainage proceeds can be increased by the action of certain bacteria.

Fig. 10.18: (a) Acid mine drainage (b) Groundwater polluted by heavy metals

10.6.7 Air Pollution

Mining usually involves the use of heavy equipment for drilling, blasting, excavation and transportation of soil, rock and ore particularly in open pit mines. Such activities can generate a lot of noise, heavy metal particles and dust that become air pollutants (Fig. 10.19a). Limestone quarries used in the production of raw material for cement manufacture in some developing countries have generated so much dust pollution that it has negatively impacted the quality of life in communities surrounding the quarries. Everywhere near these quarries is enveloped by dust that has caused respiratory diseases and turned these communities into blighted areas or "brownfields". Radioactive dust from mining of uranium is a risk to human health if unregulated (Fig. 10.19b). In the confined spaces of underground mines, different kinds of aerosols can be produced by mining activities that are harmful to human health. Coal and coal waste generally contain trace impurities of toxic metals such as arsenic, lead, mercury, nickel, vanadium, beryllium, cadmium, which can be hazardous if released into the atmosphere by substantial coal burning.

Fig. 10.19: (a) Dust from surface coal mining (b) Radioactive dust from uranium mining

Fires and gas outbursts in coal mines generate ash and smoke that are additional hazards impacting the health of miners and are capable of causing deaths in addition to environmental impact. Presence of methane is probably the most characteristic source for fires although flammable and combustible liquids are often stored underground in most mines and pose a special fire hazard. Also, it is well established that coal mining is associated with chronic health problems among miners due to inhalation of coal dust causing respiratory diseases such as black lung disease, which causes permanent scarring of the lung tissues. Studies have shown that long-term exposure to radon gas and its decay products from uranium mining and milling increases the risk of developing lung cancer.

10.6.8 Ecosystem Damage

Ecosystem is a community of living organisms that live and interact with each other in a specific environment. Ecosystems of various types (e.g. marine, terrestrial etc.) function because of the continuing exchange between the biotic (living) and abiotic (nonliving) components. Mines are highly damaging to the ecosystems surrounding them both from the physical and chemical viewpoints related to land degradation and release of toxic metals to the environment. For example, the removal of vegetation as a result of mining operations, particularly in tropical vegetated areas leads to the destruction of plant life. With reduced plant biomass, less carbon is being processed via photosynthesis, which leads to less oxygen production, less standing biomass and reduced transfer and cycling of nutrients. Soil contamination and acidification are toxic and can limit the amount of plant life and the absence of vegetation cover accelerates erosion and environmental degradation.

Review Questions

1. Briefly describe the mining cycle and highlight the role and responsibilities of a geologist in mining.
2. Attempt a classification of mining methods and explain the various criteria for selecting a mining method.
3. Describe artisanal and small-scale mining and explain why it is more prevalent in Africa. What types of mineral deposits are most amenable to artisanal mining. Give examples of minerals extracted.
4. Draw a labeled diagram of an open-pit mine and label the major features. What type of orebodies are suitable for surface mining?
5. Discuss the environmental impacts of mining, and suggest some mitigation measures.

Chapter 11: Mineral Economics

11.1 Introduction

Mineral economics is an interdisciplinary study involving the applications of economic principles in the exploration, exploitation and final use of economic minerals. Economics is basically a study of scarcity, and how scarce commodities are distributed among competing demands. Economics has significant implications on the use of resources, and the relationship between the consumption and production of mineral commodities. Metals and minerals are consumed worldwide; and the core aspects of mineral economics deals mostly with their supply and demand, and other issues such as inquiries into historical use, future markets, and substitution potential for the various mineral commodities. In recent years, mineral economics has delved into aspects of public policy and the evaluation of private and public costs as well as the socio-economic benefits of mineral resource development.

Mineral economics is a diverse and complex discipline which is a bit marginal to economic geology, but it has become important for geologists to be aware of the impacts of *geoeconomics* factors, and even *geopolitics* on the *exploration* and *exploitation* of minerals. For example, exploration companies at times have to make major decisions on what target metals they need to select for exploration, and in what part of the world, or what type of ore deposit can be prioritized for expedited mine development. Such decisions may be based on current economic conditions of global supply and demand and commodity prices and other related factors. In recent years, the expanding interests of many corporations and investors in a wide range of minerals have utilized the varied backgrounds and global experience of geologists in commodity evaluations and forecasts in making vital decisions. Many geologists are now employed in important management positions where they may need to provide advice to guide corporate decisions regarding new ventures in mineral resources exploration and exploitation worldwide.

The production of mineral commodities begins with exploration and evaluation followed by mining and processing of the ores to produce refined metals which are used by manufacturers to produce industrial and consumer goods. Industrial minerals follow a similar pathway of exploration and extraction and end up with the production of purified minerals rather than metals, hence the usage of the term "metals and minerals ".

11.1.1 Economic Importance of Metals and Minerals

As described earlier in Chapter 1, metals and industrial minerals are of considerable economic importance as sources of raw materials for industrial development and growth of the economy of many nations. Mineral commodities, such as aluminum, copper, nickel, steel, and zinc, form the basis of every modern industrial economy and are vital for economic growth, and improving the quality of life. For example, metals and minerals provide indispensable raw materials for major industrial sectors ranging from construction, transportation and infrastructure, to communication and national defense, and improving the overall functioning of modern society. Minerals also generate substantial revenues for national economies. For example, in 2020, $82. billion worth of nonfuel minerals was produced by U.S. mines, out of which metals contributed $27 million, and industrial minerals $55 million. Thus, the industrial minerals have outpaced the metals in terms of economic benefit to the national economy.

Metals and minerals are being used in larger quantities than ever before and in an increasingly diverse range of applications and their demand is likely to continue to increase in the near future. With the increase in demand has been an increase in concern regarding the reliability of supply for these and other mineral commodities because production of some critical metals has become concentrated in relatively few countries (for example, tungsten, rare-earth elements, and antimony in China; niobium in Brazil; and platinum-group elements in South Africa and Russia), thus increasing the risk for supply disruption owing to political, social, or other factors. Also, the awareness of impacts of mining on human health and environment has affected production and ultimate supply of some critical metals. These concerns have led a number of countries to attempt to identify those mineral commodities that are viewed as most "critical" to the national economy and (or) security if supplies should be curtailed while some countries have identified certain minerals as "strategic" to economic development and need to be prioritized for development.

Although metals and industrial minerals have geological similarities with respect to their mode of occurrence and extraction, there are significant differences in their economic behavior. Compared to metals, industrial minerals are very bulky and characterized by high volume and low unit prices. Wherever industrial minerals are mined they are amenable to large volume extraction, and supply is often steady and price fluctuations are minimal. In contrast to metals production, where price is generally the critical variable, industrial minerals tend to be more sensitive to volume considerations. These differences affect how they are traded in the global market.

11.2 Categories of Metals and Minerals

From the economic viewpoint, metals can be classified into two categories; *precious metals* and *industrial metals*.

11.2.1 Precious and Industrial Metals

Precious metals include gold, platinum, and silver. They are rare and can have a high economic value associated with them. Precious metals also derive their value from what they're used for: for example, a metal such as gold is used not only for jewelry but also considered a safe-haven investment in times of economic uncertainty. Some precious metals are used in industrial and manufacturing processes including gold, platinum and palladium, - a metal in the platinum group that is used in electronics. Industrial or base metals which include copper, aluminum, iron/steel, and zinc. are used in the construction, manufacturing, and technology industries. A subset of the industrial metals is the so-called *"high technology metals"*.

High-tech metals are those metals that support the rapidly growing high-technology industries, which are fueled by consumer demand for a high-tech, environmentally sustainable future. These high-tech devices range from consumer electronics, medical devices to advanced defense systems. High-tech metals are often rare and very expensive as they may be hard to find, difficult to extract in economic quantities, or have unique properties that make them difficult to substitute with lower cost metals. Examples include the group called the rare earth elements (REE) which has over 20 elements, and other designated metals such as cobalt, tantalum, lithium, gallium, niobium, selenium and zirconium.

11.2.2 Critical and Strategic Minerals

Metals and minerals have also been designated by national governments as *"critical"* or *"strategic"* based on their relative importance to the nation's economy and national security.

Critical minerals are mineral commodities that have important uses and no viable substitutes for which a threat to supply could create harm to the economy. They normally are minerals that occur only in small quantities and the supply chain is considered fragile. For example, the United States has essentially no reserves of rare earths, columbium (niobium), strontium, manganese, tantalum, or cobalt, sheet mica or bauxite ore. Since these minerals are important in various industrial processes, they are regarded as critical and are given priority in exploration and mining. The United States has designated 23 minerals including lithium (Li), manganese (Mn), niobium (Nb), platinum-group elements (PGE), rare-earth elements (REE), rhenium (Re), selenium (Se), tantalum (Ta), tellurium (Te), tin (Sn), titanium (Ti), vanadium (V), and zirconium (Zr). The European Union in 2020 also identified 30 mineral raw materials as critical, based on their importance in the economy and supply risk.

Strategic minerals are those minerals that are of vital importance to a country's national defense, energy and industry sectors, and are prioritized for exploration and development. Strategic minerals do not entail national security or supply risks. For example, the United States has identified certain minerals that are used in military equipment, jet engines, nuclear plants and space research as strategic minerals, e.g., manganese, titanium, zirconium and yttrium. Nigeria has identified seven mineral commodities as strategic minerals; iron ore, lead-zinc, gold, coal, bitumen, barite and limestone. They are prioritized for development to support iron and steel, and cement production.

11.3 Economic Theory of Supply and Demand

In order to understand the economics of production and consumption of minerals, it is very relevant as a starting point particularly for geologists with little academic knowledge of economics, to describe the basic economic *theory of supply and demand*. This theory is based on two separate "laws," the law of demand and the law of supply. The two laws interact to determine the actual market price of products or commodities and the volume available for sale in the market. The resulting price is referred to as the equilibrium price and represents an agreement between producers and consumers of a commodity. In equilibrium, the quantity of a good supplied by producers equals the quantity demanded by consumers. However, several independent factors can affect the shape of market supply and demand, thereby influencing both the price and quantities that occur in markets. These principles also apply to the demand and supply of minerals in a global market.

11.3.1 Law of Demand

The law of demand states that, assuming all other factors remain equal, the higher the price of a good, the fewer people will demand that good. In other words, the higher the price, the lower the quantity demanded. Consumers will purchase less of a product at a higher price because as the price increases so does the consumer consider other alternative choices (opportunity cost).
In theory, the quantity of a commodity demanded depends solely on the price of that commodity. However, in reality, there are usually other factors (non-price) at play, such as the prices of other commodities, the incomes and preferences of consumers, and seasonal effects. When the price-quantity combinations are plotted on a curve, it is known as a demand curve, with price represented on the vertical axis and quantity represented on the horizontal axis (Fig. 11.1). A demand curve is almost always downward-sloping, reflecting the willingness of consumers to purchase more of the commodity at lower price levels. Any change in non-price factors would cause a shift in the demand curve (Fig. 11.2), whereas changes in the price of the commodity can be traced along a fixed demand curve.

11.3.2 Law of Supply

The law of supply involves the relationship between various prices and the quantity potentially offered by producers at each price, assuming all other factors remain constant. When the price-quantity relationship is plotted on a curve with price represented on the vertical axis and quantity on the horizontal axis, it is known as a supply curve (Fig. 11.1b). A supply curve is usually upward-sloping, which means that the higher the price, the higher the quantity supplied. Producers will supply more at a higher price because the higher selling price justifies the higher opportunity cost of each additional unit sold. In practice, apart from price, there are potentially many other factors that affect the quantity supplied, such as the prices of substitute products, the production technology, and the availability and cost of hiring workers, and capital and transportation costs. Like the demand curve, any change in non-price factors would cause a shift in the supply curve (Fig. 11.2b). whereas changes in the price of the commodity can be traced along a fixed supply curve.

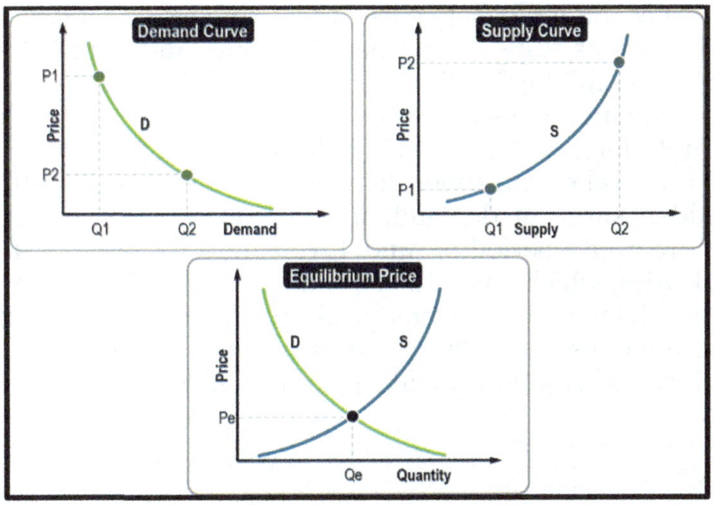

Fig. 11.1: Demand and Supply curves and Equilibrium price

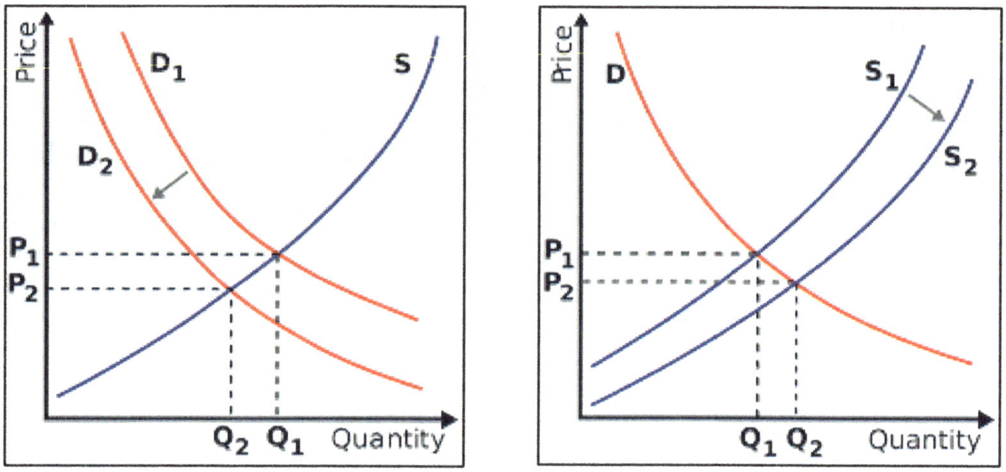

Fig. 11.2: (a) Shift in Demand (b) Shift in Supply

11.3.3 Market Equilibrium and Price Mechanism

In economic theory, the market price of any commodity is established in the intersection of supply and demand curves (Fig. 11.1c). This is the equilibrium point where the quantity demanded is equal to the quantity supplied. In an economy free of restrictions, supply and demand of commodities are usually balanced through the price mechanism. Generally, as price increases, producers are willing to supply more, while consumers will demand less. The opposite applies when the price falls.

If for any reason buyers want to purchase more of a product than is available at the prevailing price, they will tend to offer a higher price which will induce more supply. If buyers decide to purchase less than is available at the prevailing price, suppliers can lower the price to attract increased demand. Thus, the price mechanism plays an important function by determining what quantities of a commodity are to be produced as well as determining which goods are to be produced, how they are to be produced, and how they will be distributed (or who gets to buy them).

However, in practice, multiple factors can affect both supply and demand, causing them to increase or decrease in various ways. Variations in supply and demand can also lead to price changes. All the above possibilities indicate the varied role of price in maintaining market equilibrium. Based on economic theory and the assumption that the market is competitive and free from government interference, we can conclude that the scale of mineral production will be determined mostly by the level of consumer demands and the cost of getting the commodities to them. If there's a sudden surge in demand, it will lead to price increase which will induce producers to supply more over a period of time. Inversely, a decrease in demand will lead to the lowering of the market price, which will induce producers to produce less. The price system thus generates an effective mechanism for consumers and producers to make economic decisions. The market for metals and minerals generally responds in the ways described above except there is often a time lag for actions and reactions to take place in a global environment.

11.4 Economic Analysis of Global Demand and Supply

Mineral commodities are vital for economic growth and improving the quality of life. They are involved in trade among countries of the world, therefore demand and supply of minerals are global in scope. Demand for minerals originates from the appetite or needs of consumers for industrial and consumer products that lead to purchases of mineral commodities for construction and manufacturing of diverse products for various sectors of the economy. However, before a mineral can attain any value, it needs to be located, extracted, and converted into commodities for which a demand exists, and delivered to consumers in the required quantities and qualities at the needed times and locations. Minerals have been used since ancient times, and continued to be used in larger quantities than ever before for an increasingly diverse range of applications.

11.4.1 Global Demand of Mineral Commodities

Global demand for minerals is likely to be affected by three general factors: (1) the uses for mineral commodities, (1) the growth of population that will consume these mineral commodities, and (3) the standard of living of consumers that will determine the quantity that each individual consumes. Price of mineral commodities is not a major factor in demand if there is a need for the consumer products. As a matter of fact, for precious metals such as gold, increased prices may result in higher demand because of increased value for the affluent consumers.

As new materials and applications are found, markets for mineral commodities can expand considerably. Within the last few decades there has been a surge in demand in the area of high-tech consumer electronics (mobile phones), electric vehicles, green technologies, and aerospace and defense systems. A good example of new uses for mineral commodities is the surge in demand for the rare earth elements (REEs) that have in the past decade become critical components in high tech devices, medical devices, semiconductor chips and green energy transformations (solar, electric cars etc.). This has created an increased demand for supplies of REEs and other high-tech metals since the early part of this millennium.

Annual demand doubled to 125,000 tons within the last decade and is projected to reach 325,000 tons by 2030, driven by increasing use in green technologies and advanced electronics that is creating enormous pressure on global production. As shown in Figure 11.3a, the increased global demand has been met by a similar increase in supply. The main uses of REE are for producing magnets and catalysis that are important in green energy and high efficiency electric generators. This has also led to volatility in commodity price, indicating that global supply has had difficulty keeping pace with the growth in demand (Fig. 11.3). Moreover, China is the dominant producer (70% of global production) and has once threatened to cut supply which led to a spike in price.

Another example of how uses can have positive or negative effect on demand is the element strontium, which was not widely used for industrial application until the 1960s, when it was found to be the most cost-effective means of preventing radiation from escaping from color television picture tubes. Production increased from less than 50,000 tons in 1870 to over 500,000 tons in 2006 after which strontium production collapsed and within a few years, in 2010, production was nearly halved (Fig. 11.4). This collapse in supply was attributed to the introduction of today's LCD flat screens because the classical cathode-ray tube contained lead and X-ray radiation was no longer in use as a protection from X-ray radiation.

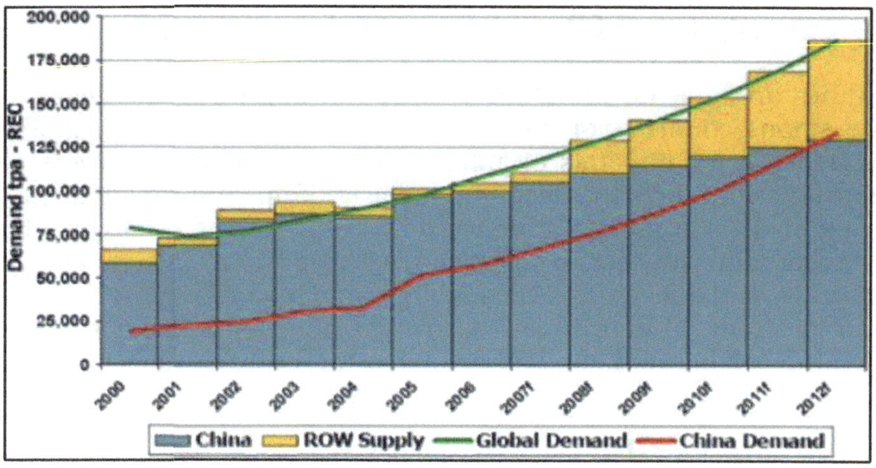

Figure 11.3: Demand and Supply of REE by China

Growth in world population, together with improvements in standards of living in many countries has fueled the demand for mineral commodities. World population has experienced exponential growth and doubled to 6:1 billion from 1960 to present most of it in the developing countries. For example, the population of China has grown by almost 1 billion in the past 60 years. Demand continues to increase as a result of population growth and better standards of living in developing countries that are now able to improve their economies and increase consumption of new products. Demand for minerals stimulates their supply and emerging economies such as China, India and the United Arab Emirates have been key drivers of global growth and demand.

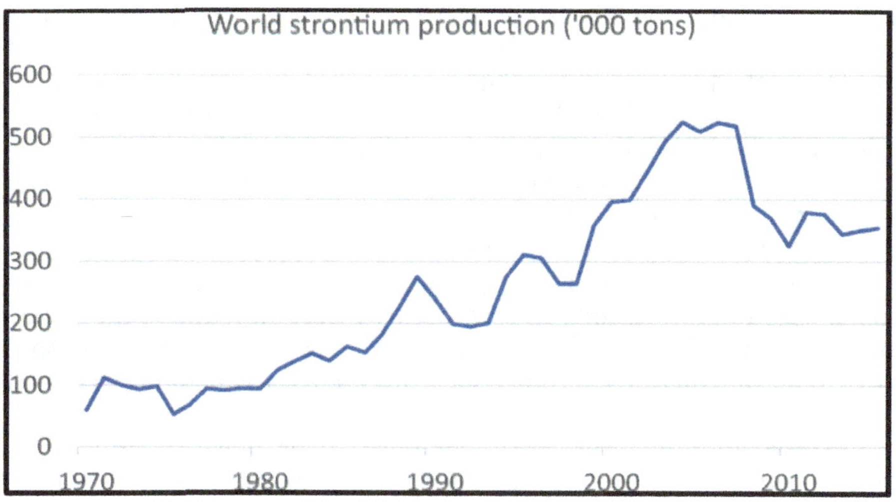

Figure 11.4: World demand for Strontium

For example, in China since the late 1980s, economic growth has been between 7 and 18 percent annually, with an average of 9.2% between 1990 and 2021. China has been undergoing industrialization, moving through a series of stages that include development of infrastructure such as buildings and roads, followed by development of light manufacturing involving more use of copper, followed by development of heavy manufacture that increased the consumption of aluminum, iron ore, and steel. As shown in Figure 11.5(a), China started experiencing the era of the so-called "steel intensity" in the early 2000s as demand was driven by a sustained period of rapid industrialization and urbanization, requiring high levels of investment in infrastructure, buildings and machinery. Demand seemed to have leveled up around 2015 when China transitioned into the consumer products stage with increased consumption of durable goods such as automobiles and electronics products which increased the demand for metals with specialty applications such as nickel, which is used in stainless steel, and high-tech metals such as the rare earths (Fig. 11.5b).

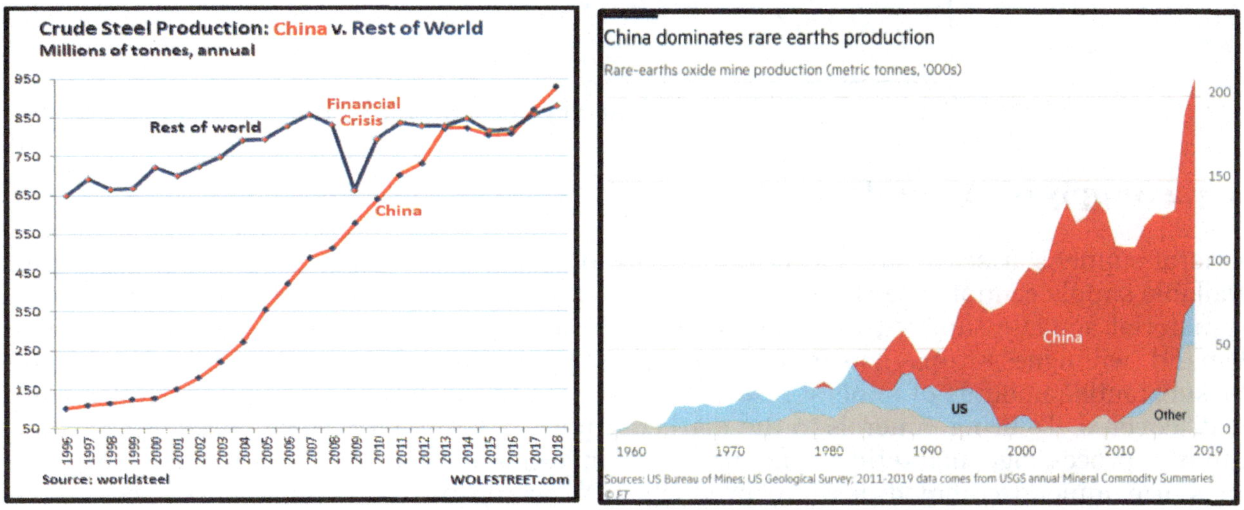

Figure 11.5 (a) China's Steel production (b) China's RRE production 1960-2019

In accordance with economic theory, a change in demand would normally affect price levels, and immediately cause the quantity supplied to change in response. Consequently, shortages and surpluses of commodities and the associated price fluctuations should be very short-lived. However, in reality, these theoretical ideals may not happen for mineral commodities. The demand for, and prices of, many metals are somewhat highly volatile, and can fluctuate a lot over a short time span (Fig. 11.6). The volatility in prices occurs because the market is not perfectly competitive because there is always a time lag for suppliers to react quickly to changes in demand. Converting mineral reserves to commodities for the markets may take up to four years, so shortages can persist and lead to price hikes. In exceptional situations suppliers may have a stockpile of low-grade slag or mine waste from which metals can be quickly recovered to meet change in demand. Mineral commodities unlike manufactured products do not have unique differences that can be selling points to consumers' taste. Metals are "bland" products that are traded in a global marketplace through bilateral agreement or in metal exchanges, and small fluctuations in demand (shortfalls or surpluses) can quickly affect prices, causing volatility.

Global demand for mineral commodities is expected to continue to increase due to the exponential growth of world population, increased standard of living and development in countries of Africa and SE Asia and the development of new technologies and products will lead to the use of a greater number of mineral commodities in increasing quantities.

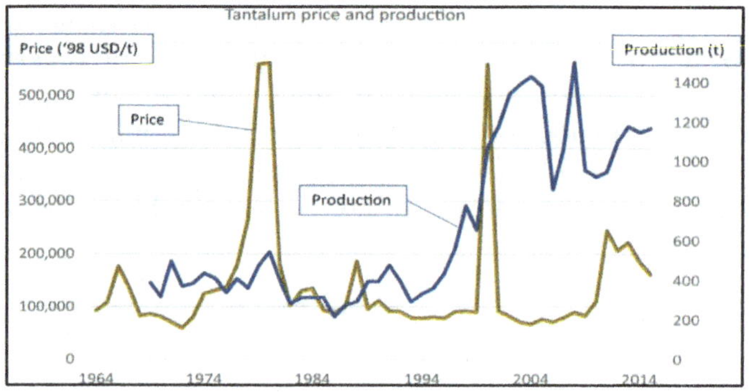

Figure 11.6: Volatility of metal price and production

11.4.2 Supply of Minerals

Mineral supply starts with the physical existence of economic minerals in Nature, and the total available supply cannot exceed the mineral's known occurrence in the Earth's crust. The amount of material actually supplied to society (economic supply) is that quantity generated by actual demand (willingness-to-pay), as may be moderated by the cost of production and related factors. So, substantial amounts of mineral resources exist deep in the ground untapped with the need to be discovered. To bring minerals to the market requires a lot of effort in exploration, evaluation, mining, processing, marketing and financial investment. Industrial activities extract and transform mineral commodities into products that people buy and use. In many cases, these activities come with direct or accumulative environmental consequences that can pose serious threats to ecosystems and human health, thereby creating potential restrictions on mineral production. For example, uranium is a radioactive element, and its production has faced restrictions due to potential damage to the environment.

In economic theory, we learn that supply occurs in response to current and anticipated demand, and when demand increases, prices tend to rise and/or producers deliver more goods to the market to keep up with the rising demand. In the global mineral market, when demand increases faster than supply, temporary shortages can lead to increase in price. When supply increases faster than demand, excess supply in the form of stockpiles can occur which can lead to a fall in price until the excess is absorbed. However, it is generally believed that the supply of most mineral commodities is inexhaustible because the quantity of minerals available at each ore deposit occurs at various levels of availability which are sometimes classified as "resources" and "reserves". Resources and reserves are often reported by the global mining industry to indicate the quantity and quality of metal or minerals available as potential supply (Fig. 11.7).

Resources are defined as mineral concentrations of sufficient size and grade and in such a form and amount that economic extraction of a commodity from the concentration is currently or potentially feasible. *Reserves* refers to Identified resources that meet specified minimum physical and chemical criteria related to current mining and production practices, and that can be economically extracted at today's prices and today's technology. Many references to the supply of mineral commodities include only the "reserves" whereas the reserves represent only a subset of the potential mineral endowment of a mineralized area, with resources forming a larger proportion of the less well-quantified mineralized material present within each mineral deposit. However, as the ore deposit is exploited, known reserves may diminish with time, but further exploration may identify more reserves from identified resources. This may be accompanied by the delineation of more resources from the surrounding poorly delineated zone of mineralization.

The dynamic relationship among these components of a mineral deposit often guarantees more production beyond the earlier identified reserves, and thus extend the projected life span of a mine.

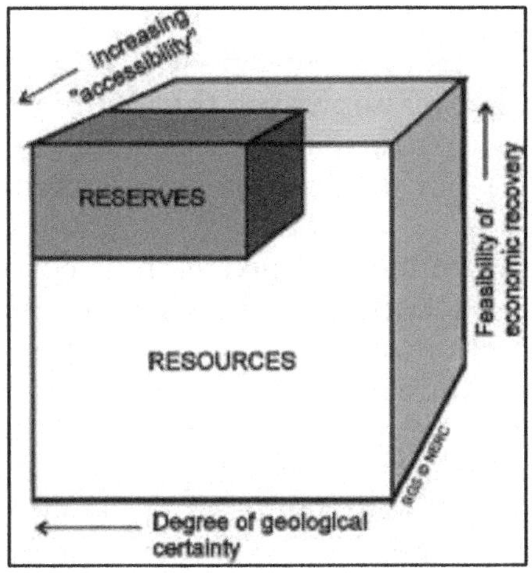

Fig. 11.7: Reserves as part of Resources

Reserves can further be classified into *inferred, indicated* or *prove*n which identifies the level of certainty in the estimates of mineral quantity and quality. It is the *proven* reserves that are immediately recoverable to meet increase in production in response to an increase in demand. However, it still takes time, sometimes up to 4 years to get all the factors of production in place to convert reserves in the ground into marketable commodities for mineral supply. Based on current estimates, global mineral reserves are 20 to almost 1,000 times larger than present annual production of several metals. This is an indication that estimates of global mineral reserves are large and therefore adequate to supply world mineral demand for most mineral commodities in the next 50 years. There are some well-known factors that could affect the future of mineral supplies, and they include *geopolitics*, wars, social constraints, insecurity, laws, environmental regulations, land-use restrictions, *geoeconomics*, and infrastructure.

11.5 Factors Affecting Future Availability of Minerals

The future availability of minerals is a major concern to policy makers because minerals are non-renewable, and human appetite for consumption is insatiable. There are several factors that can affect the future global supply of minerals, and they include geologic, technological, economic, environmental, social and political,

Geological factors are concerned with the presence of mineral deposits, - their depths, grade, tonnage, and geometry and the ease of making new discoveries by existing mineral exploration techniques. Other issues relate to the ease of extraction and recovery of metals and other commodities. Exploration activities continue on a global basis, although new discoveries are becoming more difficult and potential locations are becoming more remote. Discoveries are the most important stimulus to exploration because they provide a reason to explore areas or geologic environments that were considered previously to be without potential. Discoveries range from recognition of a new, completely unfamiliar type of deposit, through realization that a familiar deposit type might be found in a new area, to recognition of possible lateral or depth extensions of known deposits or districts. Nonetheless, finding new mineral deposits is becoming more difficult and more expensive as the more easily located deposits have been found and the potential new deposits now lie at greater depths in the crust or the most remote areas of the world. It is estimated that about 75% of areas yet to be explored that are likely to contain new minerals are located in the jungles of Africa.

Technology and productivity factors can greatly influence the quantity of mineral materials that is supplied to the market by improving the ability to find, extract, and process the mineral resources. Technological changes through new machinery, improved processes and the use of computer control systems and labor productivity may contribute to increasing productivity in mineral production and supply.

Economic factors and costs of production such as in discovering new mineral deposits; to extract the minerals; and to process, concentrate, and purify the minerals play a significant role in shaping mineral supply. Some countries are able to produce their commodities at a lower cost than others, either through natural endowment of better-quality deposits or lower production costs by factors such as lower energy, skilled labor, and taxes. Mineral commodities are heavily traded in the global market so that lower cost sources displace production in higher cost areas.

*Environmenta*l factors relate to the attributes of the environment in which the minerals are mined or processed, such as water and air quality, endangered species and scenic beauty. There is an increasing tendency that to protect human health and environment may lead to government regulations that will either slow down or restrict land access to the development of new mineral deposits in mining companies.

Political factors at local, national, and international levels can affect mineral supply particularly in relation to political stability, predictability of laws, political corruption, the independence of the judiciary, the limits on litigation, excessive taxation, the protection of land tenure, the willingness of the host country to allow or facilitate development of the resource and repatriation of profits, These conditions are particularly applicable to developing countries in Africa and Asia where military coups can destabilize legitimate governments.

Social and cultural factors may affect mineral supply as a result of community acceptance of resource development, inter-tribal feud for control of resources that occurs in some African countries, lack of security for workers and fixed installations, armed banditry, poor infrastructures and shortage of skilled labor. All these factors can come together to inhibit mineral production.

The factors analyzed above do not operate separately, but rather as parts of an integrated system, which definitely discourages exploration and mining. The development of globalization of mineral supply is a positive factor that ensures that the world mineral markets can respond collectively to change in demand. Economic drivers such as the impact of globalization, technological improvements, productivity increases, and efficient materials usage are at work simultaneously to impact minerals markets and supply, so that the historical price trend of mineral prices that has been volatile a times, remains steady over a long time. Thus, the price mechanism continues to operate efficiently in ensuring that adequate supply meets market demand.

Recycling of metals has become a significant factor in the supply of many of the metals used in our society, such as steel, copper and aluminum. Recycling provides enormous environmental benefits in terms of energy conservation, reduced volumes of waste, and reduced emissions associated with energy savings. For commodities that are recycled, rising prices tend to bring additional scrap to the marketplace, such as in the case of silver where significantly higher prices can bring large amounts of material to the market. Currently metal scraps of iron, copper, aluminum and nickel are very active in the metal market. Gold is also recycled by the melting of old or discarded jewelry.

11.6 Global Production and Consumption of Selected Metals

Metals and minerals of global economic importance are categorized into three groups: (1) precious metals (2) industrial metals and (3) high-technology metals (Table 11.1), The global supply and demand and uses of these essential minerals are described in the following section to assist in the understanding of the economic drivers of supply and demand in the global marketplace

11.6.1 Precious Metals

Supply of the precious metals comes to the market from mining production and recycling of scrap and obsolete material. Mining accounts for between 70-85% of the total new supply, depending on the metal, with recycled material accounting for the remainder. The proportion fluctuates with time, reflecting the changing cost of production, reclamation values and the economic outlook, amongst other factors. Demand for gold and silver derives from the jewelry industry and for the purpose of investment. Many banks and individuals store their reserves or wealth in gold and silver. These gold or silver bars are stored in vaults. And the stock of stored metal in these vaults that has accumulated over many years are also available as supply to the market.

Gold

Since the times of the Egyptian Pharaohs 3500 years ago, gold has been a commodity of value worldwide for use in jewelry and adornments, and in the making of coins and other financial assets by banks and investors. Small amount of gold is used in technology applications (computers and electronics). Gold's diverse uses, in jewelry, technology and by central banks and investors, mean different sectors of the gold market rise to prominence at different points in the global economic cycle. However, jewelry use is the primary physical use of gold, with the amount used exceeding 2/3 of annual production, Gold is in very high demand worldwide, and global production in 2020 was estimated at 3250 metric tons with China as the leading producer with 380 metric tons (12%). Other major producers are Australia (10%), Russia (9%), USA (6%) and Canada (5%).

Table 11.1: Categories of Metals in Global Trade

	Precious Metals	**Industrial Metals**	**High-Tech Metals**
	Gold (Au)	Copper (Cu)	Rare Earths (REE)
	Platinum (Pt)	Aluminum (Al)	Lithium (li)
	Palladium (Pd)	Iron/Steel (Fe)	Tantalum (Ta)
	Silver (Ag)	Nickel (Ni)	Niobium (Nb)
	*******************	Zinc (Zn)	Zirconium (Zr)
	*******************	Lead (Pb)	Gallium (Ga)
	*******************	Tin (Sn)	Hafnium (Hf)

Africa is a leading continent in gold output, contributing over 16% of world production in 2019. Prior to 2007, South Africa was the global leader in gold production, but output has fallen steadily over the years from 400 metric tons in 2000 to 120 tons in 2020. Ghana is now Africa's leading producer with annual production of 140 metric tons. African countries with estimated 20% of global reserves export over 95% of their gold output mostly to Europe, Middle East and Asia. Mine production accounts for about 75% of annual supply while the rest is derived from metal recycling particularly of used jewelry. The world's leading consumers of gold are India, China, United States and United Arab Emirates. India and China accounted for 57% of gold jewelry consumption globally. Gold price in 2020 was about an average $2000 per ounce.

Silver

Silver, the "white metal," is highly sought after by investors and industrialists alike. Producers of electronics and medical supplies appreciate silver for its atomic properties. Collectors of silver coins and bars and jewelers cherish silver's unmatched shine and malleability. Silver is also used in making tableware, mirrors, in dental alloys, solder and brazing alloys, electrical contacts, solar panels and batteries. Global supply of silver in 2020 was estimated at 27,000 metric tons with a value in the order of $17 billion. The leading global producers are Mexico (6000 tons), Peru (4000 tons), China (3570 tons, Russia (2000 tons and Poland (1500 tons). Silver is often a co-product of the mine production of other minerals, such as gold or copper. World's leading consumers of silver are India, United States, United Kingdom, Canada, Germany, Australia and China in that order of importance. Current price of silver is about $25 per oz.

Platinum and Palladium

Platinum and palladium are the two most important metals of the platinum group. Both metals have many industrial applications including use in automotive catalytic converters, electrodes in medical equipment, in computer hard disks, mobile phones and in fine jewelry. Platinum is also used in dentistry, in the form of surgical instruments and electrical contacts, as well as in the production of strong, permanent magnets. Platinum alloys are used in capsules for radioactive power sources. Global mine production in 2020 of platinum was 170 tons and palladium 210 tons. Africa dominates the world's production of platinum and palladium with a major contribution of 130 tons of platinum from South Africa (76% of world production) and Zimbabwe 15 tons (9%). South Africa also contains 90% of the world's reserves estimated as 70,000 tons. Other platinum producing countries are Russia with 21 tons, Botswana, Tanzania, and Madagascar.

Palladium supply averages 200 tons annually of which Russia surpasses South Africa with 90 tons, South Africa, 70 tons, Canada 20 tons, United States 14 tons and Zimbabwe 12 tons. South Africa is the only place in the world where platinum ores are mined directly from rocks. Elsewhere, most platinum is mined as a by-product of the refining of other metals, mostly nickel. Recycled platinum from auto-catalysts and jewelry averages about 50 metric tons annually. In 2020, China was the world's largest platinum consumer, with a consumption of approximately 57 metric tons, followed closely by Europe with a platinum consumption of around 51 metric tons. For palladium, China accounted for the largest consumption at 31 percent followed by North America which was previously the largest consumer of palladium. The prices of platinum and palladium have fluctuated over time, but platinum price has always been higher until recently when it was surpassed by palladium. Currently in 2022, palladium is the most expensive of the precious metals with a price of $3000 compared to $$1200 for platinum.

11.6.2 Industrial Metals

Industrial metals are commodities that are widely used in the manufacturing industries and other related applications. Copper is currently the most important while aluminum is the most widely used metal. Iron (and steel) is the most voluminous of the industrial metals because of its widespread application in construction and production of heavy machinery. Other important industrial metals include tin, nickel and zinc. Demand for industrial metals has increased substantially in the last two decades due to consumption by Chinese manufacturers (Fig. 11.6a). Supply of these metals come mostly from recycling with mining production contributing from 1-30% (Table 11.2). In 2020, Supply has also shown an upward trend over the years, but with a bit of fluctuation in the last two decades following a similar pattern in demand (Fig. 11.6b). The main price driver for base metals was the rebound of Chinese industrial production as the country consumes 50% of the world's base metal production. Chinese demand remains strong, indeed its strongest level since 2011.

Copper

Copper (Cu) is the most important industrial metal because of its widespread applications by the wire and cable industry, taking advantage of its properties such as the electrical conductivity, corrosion resistance, and thermal conductivity. Demand for copper by source are: electrical and electronic products (36%), building construction (31%), transportation equipment (11%), consumer products (10%), and industrial machinery (9%). Global mine production of copper in 2020 was about 20 million tons of which the five largest producers were Chile (5.6 million tons) (Mt), Peru (2.4 Mt), China (1.6 Mt), Democratic Republic of Congo (1.3 Mt) and the United States (1.3 Mt). Recycled copper contributes about 2 million tons annually to world supply. Global demand for copper is estimated as 25 million tons annually, and China is the leading consumer with 54% of the volume. Copper supply has increased over the last 10 years in response to China's industrialization and to meet China's demand, world supply from copper mines doubled over the period 1994-2014. The future is bright with demand expected to continue to increase as urbanization and green technologies expand.

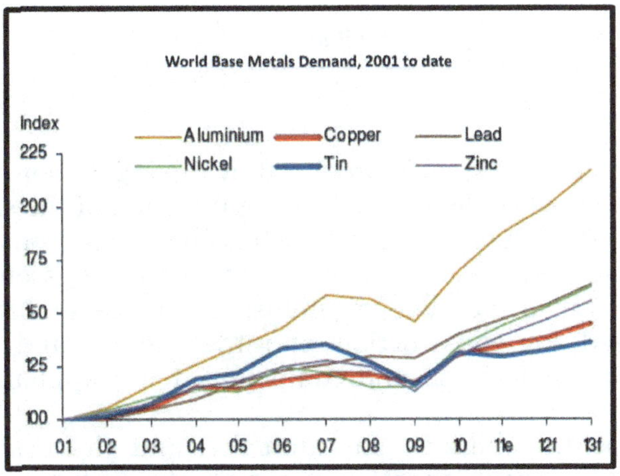

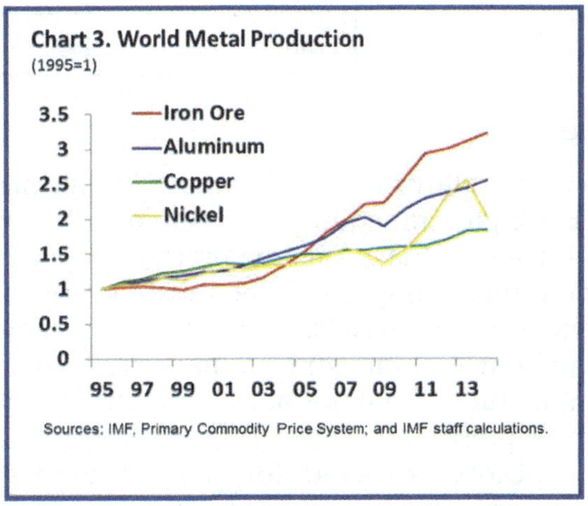

Figure 11.6: (a) World Industrial Metals Demand (b) World Industrial Metal Production

Aluminum

Aluminum is the most widely used metal after iron. It has a low density, is corrosion resistant, malleable and easily worked and a good conductor of both heat and electricity. Aluminum is produced through the electrolysis of bauxite (aluminum ore). Pure aluminum is used mainly by the electronics industry for capacitor foil, hard disc drives and conductor tracks on silicon chips. However, when alloyed with small amounts of other metals such as copper, zinc, magnesium and with silicon, aluminum becomes stronger than steel and ideal for use in aircraft, automobiles, beverage containers, and important in transportation and building industries. Global supply of aluminum from mining in 2020 was 65 million tons of which China was the leading producer with 37 million tons followed by India, Russia and Canada. Australia and Guinea are the leading primary producers of bauxite - the ore of aluminum, most of which is exported to the USA and Canada for processing because of the high energy use. Recycling contributes about 30 million tons to the world supply of aluminum. Global demand for aluminum has shown a trend that is increasing while supply has lagged behind. Price has been volatile but shows an upward trend.

Table 11.2: Industrial Metal Supply (2020 statistics)

Metal	Quantity Mining (tons)	Recycled Metal (tons)
Copper	20.7 million	2 million
Aluminum	65.2 million	30 million
Iron/Steel	1877.5 million	500 million
Nickel	2.7 million	0.2 million
Tin	0.3 million	-
Zinc	12.3 million	5 million
Lead	4.7 million	7 million

Steel

Steel is an alloy of iron, carbon and other metals. Primary steel is produced by mixing carbon and iron at high temperatures (>2600 deg. F), and is called *carbon steel*. The composition of the steel is adjusted so that it has the precise properties needed. Stainless steel which is an alloy of iron and chromium or tungsten is the most durable and most widely used steel. Steel is the most essential engineering materials, and is the foundation for a number of industries in general and manufacturing industry in particular. It is used for construction of infrastructure and the manufacture of automobiles, ships, railways, heavy machinery and equipment, tools, appliances and weaponry.

Global demand for steel is estimated at 1.85 billion tons annually, and the leading consumer is China with annual demand of 940 million tons, slightly above 50% of world demand. Other leading consumers are the manufacturing centers of Europe, USA, Mexico and Canada. China is also the world's leading producer of steel which makes a lot of impact on the global steel market. Global production of steel in 2020 was about 1.9 billion tons of which China produced 1.0

billion tons accounting for 57% of world supply. Other leading suppliers are the European Union, India, Japan and the USA. Recycling of steel scrap contributes about 500 million tons annually to world supply. Global demand for steel is rising and almost overwhelming available supply while prices are rising sharply. This trend is due to the boom in manufacturing and the construction of infrastructures worldwide including in developing countries of Africa and Asia.

Nickel

Nickel is a good conductor of heat and electricity, and is used as an alloy in stainless steel and also as a component of batteries. It is also used in gas turbines and rocket engines because of its resistance to heat. Global production of nickel in 2020 was 2.7 million tons, and the leading producer was Indonesia(305), Philippines (15%), Russia, New Caledonia and Canada. Global demand was approximately 2.5 million tons, mostly from China which accounts for about 60% of total consumption. About 70% of the demand is for stainless steel production, but the demand from battery manufacturers is projected to surge with the future production of electric vehicles.

11.6.3 High Technology Metals

High technology metals are relatively rare metals that are used in high technology devices. The variety of products using high-tech metals are almost endless: from tiny mobile phone parts through to medical applications such as hip replacements and pacemakers; from storing solar energy at home to electric vehicle components and parts for huge wind turbines, aircraft and space equipment (Fig 11.7). The list of metals that belong to this group are debatable, but some of the more commonly agreed metals are the rare earth metals, lithium, tantalum, niobium, gallium, selenium and zirconium. In the past 20 years, the world has seen the emergence and rapid growth of high technology applications in consumer electronics (mobile phones and tablets), transportation (electric vehicles), aerospace (satellites) and renewable energy (solar, wind, battery storage). This trend will continue to boost demand for high tech metals. Table 1.3 is a list and uses of rare earth and other high technology metals.

Rare Earth Elements

Rare earth elements (REEs) comprise about 20 chemically related elements that have recently found use in high-tech applications. They are used as components in high technology devices, including smart phones, digital cameras, computer hard disks, fluorescent and light-emitting-diode (LED) lights, flat screen televisions, computer monitors, and electronic displays. Large quantities of some REEs are used in clean energy and defense technologies. Global demand for REE in 2020 was about 125,000 and is projected to double in the next decade. The light REE Neodymium and Praseodymium are in high demand for use in products such as motors, turbines and medical devices. The global market for REE products is estimated to be up to 5 trillion US dollars annually.

Global production in 2020 was e 210,000 tons. China is the world's largest producer accounting for almost 60% of global annual production, estimated at 140,000 tons for 2020. Most of the remaining 40% is shared between the United States, Australia and Madagascar (Fig. 11.8). China remains virtually the only producer of the valued heavy REEs. The United States was once the world's largest supplier until the emergence of China in the mid-1990s. China remained the sole global supplier of REEs until 2012 when USA and Australia resumed production. The monopoly of China has created supplier risk and a threat to US national security needs for REE metals.

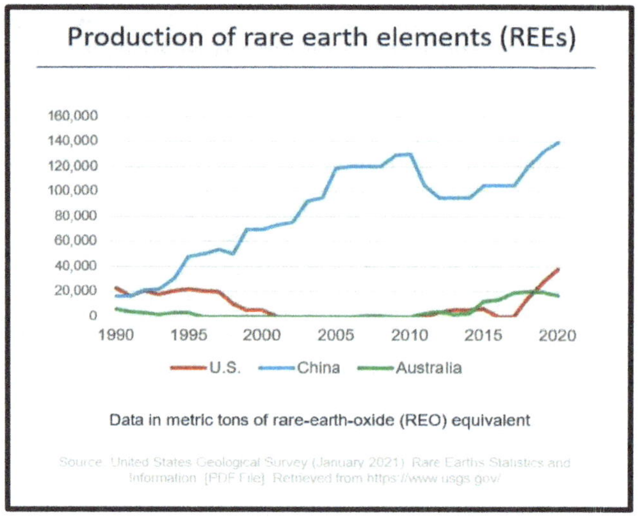

Figure 11.8: REE Production 1990 -2020

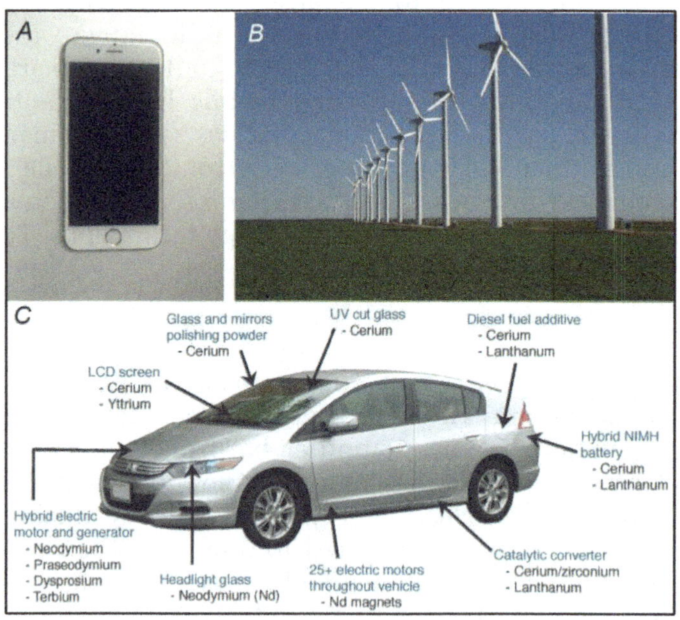

Fig. 11.7: Pictorial Illustration of Uses of High Technology Metals

Lithium

Lithium is used in rechargeable batteries for mobile phones, laptops, digital cameras and electric vehicles. Global demand for lithium exceeds 2 million tons per annum, and is expected to increase substantially in the next decade due increased production of rechargeable and storage batteries for mobile phones, solar energy and electric vehicles. Production increased considerably since the beginning of this millennium in response to global demand (Fig. 11.9a). Australia is by far the largest producer of lithium with annual production of 32,000 tons, followed by the South American countries of Chile, Bolivia and Argentina, the so-called "lithium triangle" countries. Chile, the second largest producer, has the largest world reserves of lithium. China is a major player in the lithium market because it controls 80% of the refining capacity. China is the single largest consumer of lithium because of its capacity in battery manufacturing. It imports most of its lithium, particularly from Australia.

Tantalum

Tantalum is a very hard, heavy and ductile metal that is highly resistant to corrosion. It is very conductive to heat and electricity and renowned for being nearly immune to chemical attack at room temperatures. Tantalum is considered a critical metal needed for applications in the electronics industry, in capacitors for medical appliances, mobile phones and in computer hard drives of which 35% is used by mobile phone manufacturers. Consequently, the global demand for tantalum has increased considerably since the mid-1990s. Global production of tantalum reached its peak in 2007 with over 2000 tons but has declined since then with production in 2020 estimated at 1800 metric tons (Fig. 11.9b). The major sources are Central Africa (66%), Nigeria (18%), Brazil (6%) and Australia (5%),

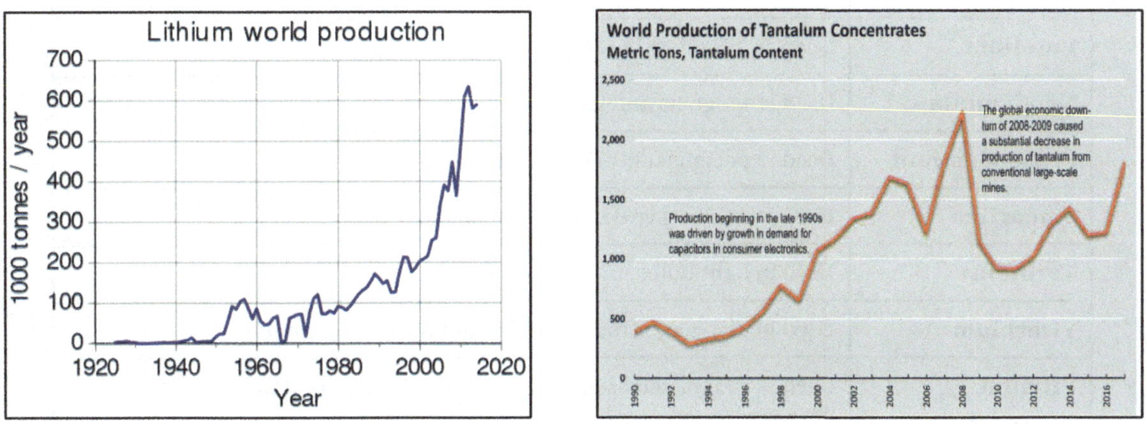

Figure 11.9 (a) Lithium world production (b) Tantalum world production

Zirconium

Zirconium is the metal derived from the mineral zircon ($ZrSiO4$), Zirconium has found widespread use in the metallurgical refractories and ceramic industries because of its heat and chemical resistance properties. However, in recent years, zirconium has found increasing application in a range of new technologies and has become an essential part of cell phones, nuclear plants, dialysis machines, paint, ceramics and catalytic converters. Most zircon is normally produced as a byproduct of the mining and processing of heavy-mineral sands where ilmenite is the principal ore mineral. World production of zircon mineral concentrates in 2020 was estimated as 1.4 million tons, of which the leading producers were Australia (550,000 tons), South Africa (350,000 tons), United States (100,000 tons), China (80,000 tons) and Mozambique (50,000 tons). World reserves of zircon are estimated at 74 million tons. The world's largest reserves are in Australia (64%), South Africa (19%) and India (5%). Demand for zirconium will continue to grow, and because of its unique physical and chemical properties it will find application in many new growth technologies, including more efficient and environmentally-friendly clean technologies.

Table 11.3: Rare Earth and High Technology Metals and Uses

Metal	Uses
Cerium	Used in light bulbs, TVs and ovens
Dysprosium	Mixed within alloys used in wind turbines,
Erbium	Used in lasers and fiber optic cables.
Europium	Used in in light bulbs, nuclear reactors and lasers
Gadolinium	Used in n magnets, nuclear reactors and magnetic resonance imaging (MRI).
Holmium	Used in magnets and nuclear reactors.
Lanthanum	Mixed within alloys that are used in batteries and hydrogen vehicles.
Lutetium	Used as a catalyst in refineries
Neodymium	Used in magnets and lasers.
Praseodymium	Used in pacemakers and guided missiles
Samarium	Used in microwave devices and magnets.
Terbium	Used in light bulbs, memory devices and x-ray
Ytterbium	Used in displays, x-ray machines and fiber-optic cable
Yttrium	Used in radars and as an additive within alloys used in high tech devices.
Scandium	Used for fuel cells and alloys used in jet planes.
Cobalt	Used in super-alloys, jet turbines and rechargeable batteries.
Lithium	Used in rechargeable and storage batteries
Tantalum	Used in capacitors, resistors and other electronic equipment.
Indium	Used in LCD screens
Gallium	Used in integrated circuits, LEDs and semiconductors.
Niobium	Used in steel alloys that are a part of jet engines and rockets.
Selenium	Used in photocells and rectifiers.
Zirconium	Used in nuclear power station

Revision Questions

1. What is mineral economics? Why is it important for geologists to have some knowledge of mineral economics?
2. Briefly describe the laws of demand and supply. When the demand for a product increase, what happens to price and supply of the product?
3. Define critical and strategic minerals. Give one reason why a country will identify minerals as strategic but not critical.
4. Write a short essay on the global demand of mineral commodities. What factors are likely to affect the future availability of metals and minerals.
5. Describe the uses, production and consumption of the following metals (a) platinum (b) rare earth elements

References

Abzalov, M (2016) Applied Mining Geology, Springer Publishers, 438pp.

Abzalov MZ, Menzel B, Wlasenko M, and Phillips J (2010) Optimisation of the grade control procedures at the Yandi iron-ore mine, Western Australia: geostatistical approach. Appl. Earth Sci 119(3):132–142

Abzalov, MZ (2008) Quality control of assay data: review of procedures for measuring and monitoring precision and accuracy. Exp Min Geol J 17(3–4): 131–144

Annels, AE (1991) Mineral deposit evaluation: a practical approach. Chapman Hall, 450 p.

Arndt, N and Ganino, C (2012) Metals and society: an introduction to economic geology. Springer, Berlin,

Bateman, AM (1950). Economic Mineral Deposits, 2nd ed. New York: Wiley, 916p.

Bates, R.L. (1969): Geology of the industrial minerals and rocks. 2nd edition (translated into Russian). Dover Publications Inc., New York, 459 pp.

Beukes, NJ, Swindell, EAJ and Wabo, H (2016) Manganese deposits of Africa. Episodes Vol. 39, no. 2, p. 285-317.

Brimhall GH, Dilles JH, Proffett JM (2006) The role of geologic mapping in mineral exploration. Soc Econ. Geol. Spec. Pub 12:221–241

Brady, BHG and Brown ET (2004) Rock mechanics for underground mining. Kluwer Academic Publishing, New York, p 628

Brown, A.C., (1992) Sediment-hosted stratiform copper deposits: Geoscience Canada, v. 19, no. 3, p. 125- 141.

Carlson, CA. (1991) Spatial distribution of ore deposits: Geology, v. 19, p. 111-114.

Carne, RC, and Cathro RJ. (1982) Sedimentary exhalative SEDEX) zinc-lead-silver deposits, northern Canadian cordillera. CIM Bull 75:66–78

Carr, o.o. (1994) Industrial minerals and rocks, 6th edition. Soc. of Mining, Metallurgy and Exploration, Littleton, Colorado, 1196 p.

Carras SN (1987) Ore reserve estimation methods and their relationship to recoverable reserves. Management and Economics of Mining Operations (Symposium Proceedings, Western Australian School of Mines, Kalgoorlie, October 1986), pp 57–67

Carter, PG (2011) Selection process for hard-rock mining. In: Darling P (Ed) SME mining engineering handbook, rd edn. Society for Mining, Metallurgy, and Exploration, Inc, Englewood, pp 357–376

Clark, GB, Mero, JL and Hustrulid, WA (2017) "Mining". *Encyclopedia Britannica*

Christidis, GE (2011) Industrial minerals: Significance and important characteristics. European Mineralogical Union Notes in Mineralogy 9(1):1-12

Ciullo, PA (1996). Industrial Minerals and Their Uses. Westwood, NJ: Noyes Publications.

Condie, KC (2001) Mantle plumes and their record in Earth history: Cambridge University Press, 320 p.

Courtillot, V., Davaille, A., Besse, J., and Stock, J. (2003) Three distinct types of hotspots in the Earth's mantle: Earth and Planetary Science Letters, v. 205, p. 295-208.

Cox, DP and Singer, DA., *eds.* (1986) Mineral deposit models: U.S. Geological Survey, Bulletin 1693, 379 p.

Dahlkamp, FJ (1993) Uranium ore deposits: Springer-Verlag, 460 p.

David, M (1977) Geostatistical ore reserve estimation. Elsevier Amsterdam, 364 p.

Davis, JG (1992) Grade control for Australian open pit old mines. In: Annels AE (Ed) Case histories and methods in mineral resource evaluation, v. 63. Geological Society Special Publication, London, pp 219–232

Dentith M, and Mudge, ST (2014) Geophysics for the mineral exploration geoscientist. Cambridge University Press, New York, 454 p.

Diehl PY, and David, M (1982) Classification of ore reserves and resources based on geostatistical methods. CIM Bull, 75, 127-135.

Dominy, SC (2010) Grade control sampling methods in underground gold mine grade control, Proc. Sampling Conf., 7–20, Melbourne,

Dunbar, WS (2012) Basic Mining and Mineral Processing. Americas School of Mines Presentation. 75pp.

Edwards, R., and Atkinson, K (198) Ore deposit geology and its influence on mineral exploration: Chapman and Hall, London, 466 p.

Ericsson, M (2007) Trends in underground mining. In: Smith M (Ed) Mining methods in underground mining. Atlas Copco Rock Drills AB, Örebro, pp 3–5

Erickson AJ, Padgett JT (2011) Geological data collection. In: Darling P (ed) SME mining engineering handbook. Society for Mining Metallurgy & Exploration, Englewood, pp 145–159

Evans, AM (1993) Ore geology and industrial minerals. An introduction. Blackwell Publishing Ltd., Oxford, 389 p

Evans, MA (2006) The mineralogy of economic deposits. In: Introduction to mineral exploration, p. 19-32. Mason, Whateley and Evans (eds), Blackwell Publishers.

Evans, MA and Moon, CJ (2006) Mineral deposits geology and models. In: Introduction to mineral exploration, p. 33-52. Mason, Whateley and Evans (eds), Blackwell Publishers.

Evans, DM, Simmonds, J and Hunt, JP (2016) An overview of nickel mineralization in Africa. Episodes Vol. 39, no. 2, p. 319-333.

Foster, RP. (2016) Mineral resources of Africa, 2008-2016. Society of Economic Geologists, Paper, 8 p.

Giuliani, G. et al. (2014) The geology and genesis of gem corundum deposits. Mineralogical Association of Canada Short Course 44, Tucson AZ, February 2014, p. 29-112

Groat, LA. (ed). (2007) The Geology of Gem Deposits. Mineralogical Association of Canada, Short Course Volume 37.

Gross, GA (1980) A classification of iron formations based on depositional environments: Canadian Mineralogist, v. 18, p. 215-222.

Groves, D.I., Goldfarb, R.J., Robert, F., and Hart, C.J (2003) Gold deposits in metamorphic belts: overview of current understanding, outstanding problems, future research, and exploration significance: Economic Geology, v. 98, p. 1-30.

Groves DI, and Bierlein, FP (2007) Geodynamic settings of mineral deposit systems. J Geol Soc Lond. 164:19–30

Guilbert, J.M., and Park, C.F Jr. (1986) The geology of ore deposits: W.H. Freeman & Co. 985 p.

Hedenquist, J.W., and Lowenstern, J (1994) The role of magmas in the formation of hydrothermal ore deposits: Nature, v. 370, p. 519-526.

Hutchinson, CS (1982) Economic deposits and their tectonic setting: John Wiley 365 p. Haggerty, S.E (198), Diamond genesis in a multiply-constrained model: Nature, v. 320, p. 34-38.

Harmer, HA and Nex, PAM (2016) Rare earth deposits of Africa. Episodes 39, no. 12 p. 381-403

Hartman HL ed. (1992). *Mining Engineering Handbook, 2nd Ed.* 2170 pp., Society for Mining, Metallurgy, and Exploration, Inc. (Littleton, USA).

Hamrin, H (1982) Choosing an underground mining method. In: Hustrulid WA (Ed) Underground mining methods handbook. AIME, New York, pp 88–112

Hamrin H (2001) Underground mining methods and applications. In: Hustrulid WA, Bullock RL (Eds) Underground mining methods: engineering fundamentals and international case studies. Society for Mining.

Hayes TS, Cox DP, Piatak NM, Seal RR, II (2015) Sediment hosted stratabound copper deposit model: U.S. Geological Survey Scientific Investigations Report 2010–5070–M, 147 p-

Hedenquist, JW and Lowenstern, JB., (1994) The role of magmas in the formation of hydrothermal ore deposits: Nature, v. 370, p. 519-526.

Harben, PW. And Kuzvart, M. (1997) Industrial minerals - Global geology. 462 pp., Industrial Minerals Information Ltd. (IMIL), Worcester Park, Surrey, UK.

Hughley, DE Chapman, GR and Bound, KA (2004) The economic importance of minerals to the UK. Brit. Geol. Surv. Publication, 42p.

Hutchinson, CS (1982) Economic deposits and their tectonic setting: John Wiley, 365 p.

James, H and Trendall, AF. 1982) Banded iron formations: distribution in time and paleoenvironmental significance, *in* Holland, H.D. and Schidlowski, M., *eds.*, Mineral deposits and the evolution of the biosphere: report of the Dalhem workshop: Springer Verlag, p. 199-218.

Jebrak, M (2006) Economic geology: Then and Now. Geoscience Canada, 33(2), 81-93.

Jensen MLY, and Bateman AM (1979) Economic mineral deposits. Wiley, New York, 593 p

Kesler, SE (1991), Mineral resources, economics, and the environment: Macmillan College Publishing Company, 391 p.

Kesler, SE (2014) Mineral supply and demand to the 21st century. Proceedings, Workshop on Deposit Modeling, Mineral Resource Assessment, and Sustainable Development

Kinnaird, JA, Bowden, P, Nex, PAM. and Milani, L (2016) Tin in Africa. Episodes 39, no. 2351- 360.

Kinnaird, JA and, Nex, PAM (2016) Uranium in Africa. Episodes 39, no. 2, 361-380.

Kirk, J., Ruiz, J, Chesley, J., Walshe, J., and England, G. 2002), A major Archean gold- and crust-forming event in the Kaapvaal Craton, South Africa: Science, v. 297, p. 1856-1858.

Kirkham, RV., Sinclair, WD., Thorpe, R., and Duke, JM., *eds*. (1995) Mineral deposit modeling: Geological Association of Canada, Special Volume 40, 798 p.

Kogel, JE, Tricedi, NC. et. al. (2006) Industrial minerals and rocks – commodities, markets and prices. Soc. Min. Metall. Explor. Publication, 1353 p.

Kuzvart M. (1984): Industrial minerals and rocks. Elsevier, Amsterdam, 454 pp. Annex 3 Kuzvart, M. (2006) Industrial Minerals and Rocks in the 21st Century

Levinson AA (1974) Introduction to exploration geochemistry. Applied Publishing, 671 p.

Lowell, JD and Guilbert, JM (197) Lateral and vertical alteration-mineralization zoning in porphyry ore deposits: Economic Geology, v. 65, p. 373-408

Marjoribanks R (2010) Geological methods in mineral exploration and mining. Springer, Berlin, 238 p

Milson J (2006) Geophysical methods. In: Moon, CA, Whateley MKG, Evans AM (eds) Introduction to mineral exploration, 2nd edn. Blackwell Publishing, Ltd., Oxford, pp 127–154.

Misra, KC (2000) Understanding mineral deposits: Kluwer Academic Publishers, 845 p.

Mitchell, AHG and Garson, MS (1981), Mineral deposits and global tectonic set- tings: Academic Press, Geology Series, London, 405 p.

Naldrett, AJ (2004) Magmatic sulfide deposits. Geology, Geochemistry and Exploration: Springer, Berlin, Heidelberg, New York, 727 p.

Neukirchen, F (2020) The World of Mineral Deposits: A Beginner's Guide to Economic Geology, Springer, 353 p.

Ohmoto, H (1996) Formation of volcanogenic massive sulfide deposits: the Kuroko perspective: Ore Geology Reviews, v. 10, p. 135-177.

Olade, MA (2020) Mineral deposits and exploration potential of Nigeria. Prescott Books, 429 p.

Olade, MA (2021) Basic economic geology and mining practices. Presscott Books, 456 p.

Peters WC (1978) Exploration and mining geology. Wiley New York, 685 p.

Pirajno, F (199) Hydrothermal mineral deposits. Principles and fundamental concepts for the exploration geologist: Springer Verlag, Berlin, 709 p.

Pirajno, F (2000) Ore deposits and mantle plumes: Kluwer Academic Publishers, Dordrecht/Boston, 556 pp.

Pirajno, F (2004), Hotspots and mantle plumes: global intraplate tectonics, magmatism and ore deposits: Mineralogy and Petrology, v. 82, p. 183-216.

Pohl, WL (2011) Economic geology: principles and practice In: Metals, minerals, coal and hydrocarbons – Introduction to formation and sustainable exploitation of mineral deposits. Blackwell Publishing Ltd., Oxford, 663 p

Prentice, JE (1990) Geology of construction materials. Chapman and Hall, London, 201

Revuelta, MB (2016) Mineral Resources: From Exploration to Sustainability Assessment. Springer Publishers, 641 p.

Robb, (2004) Introduction to ore-forming processes: Blackwell Publishers, 373 p.

Sangster, DF (1990) Mississippi Valley-type and Sedex lead-zinc deposits: a comparative examination: Canadian Institute of Mining and Metallurgy, Transactions, v. 99, p. B21-B42.

Sawkins, FJ (1984) Metal deposits in relation to plate tectonics. Springer, Berlin, 325 p Stanton RL (1972) Ore petrology. McGraw-Hill, New York, 713 p.

Simonet, C. Frisch, E. and Lasnier, B (2008) A classification of gem corundum deposits aimed towards gem exploration. Ore Geology Reviews, 34. 1-2, p. 127-133.

Skinner, BJ *ed.* (1981) Economic Geology 75th anniversary volume 1905-1980: The Economic Geology Publishing Company, 964 p.

Skinner, BJ (2005) The founders of Economic Geology: SEG Newsletter, v. 60, p. 1-4.

Skinner, BJ (2015) Mineral deposits. Encyclopedia Britannica, 15p.

Taylor, CD et al (2005) Geology and Nonfuel Mineral Deposits of Africa and the Middle East. U.S. Geological Survey Open-File Report 2005–1294-E, 246 p.

Thomas, LJ (1973) *An Introduction to Mining*, pp.436, Hicks Smith & Sons. (Sydney, Australia).

United States Geological Survey (1080) Principles of a Resource/Reserve Classification for Minerals. USGS Circular 831, 12p.

United States Geological Survey (2017) Critical mineral resources of the United States. USGS Prof. Paper 1802.

Wagner, LA et al. ((2002) Economic drivers of supply and demand. USGS Open File Report 02-335.

Glossary of Common Terms in Economic and Mining Geology

Acid mine drainage - The acidic run-off water from mines due to sulfuric acid.

Adit - A horizontal or sub-horizontal passage driven into a mine from a hill or mountain side.

Alluvial – Accumulation of sediments in a river bed.

Assay- The geochemical analysis of ores and rocks for their metal content which is expressed either in percentages (%), parts per million (ppm) or ounces per ton (oz/t) for precious metal...

Backfill - Waste material used to fill the void created by mining an orebody.

Banded Iron Formation – A sedimentary deposit consisting of alternating bands of iron minerals and beds of silica common in the Precambrian.

Beneficiation - The treatment of mined material, making it more concentrated or richer.

Blasthole - A drill hole in a mine that is filled with explosives to blast loose a quantity of rock.

Block Caving - An inexpensive method of mining in which large blocks of ore are undercut, causing the ore to break or cave under its own weight.

Borehole - Any deep or long drill-hole, usually associated with a diamond drill.

Brownfield exploration - Exploration work that's done close to or at existing mines.

Chip sample - A method of sampling a rock exposure whereby a regular series of small chips of rock is broken off along a line across the face.

Chute - An opening usually constructed of timber and equipped with a gate, through which ore is drawn from a stope into mine cars.

Concentrate – what is left after the raw ore has been ground down and the waste has been removed.

Core logging - A process of recording geological observations of drill core either on paper or computer disc.

Core or Core sample - A cylinder sample generally 1-5" in diameter drilled out of an area to determine the geologic and chemical characteristic of the rock or ore.

Crosscut - A passageway driven between the entry and its parallel air course or air courses for ventilation purposes.

Cut-off grade -The minimum metal concentration or grade of ore that is considered minable at a profit. This may change with time and changing economic considerations.

Cut-and-fill - A method of stoping in which ore is removed in slices, or lifts, and then the excavation is filled with rock or other waste material (backfill),

Decline -This is a tunnel driven from the surface towards the base of the mine to allow vehicles and equipment to go in and out of the mine.

Deposit - This is a number of ore or mineral that makes exploitation worthwhile.

Diamond drill – A drill used to collect core samples from rock.

Disseminated ore - The scattering of ore mineral uniformly throughout host rock.

Dragline - A large excavation machine used in surface mining to remove overburden (layers of rock and soil) covering a coal seam.

Dredge – A common piece of mining equipment, used to sucks up dirt and gravel from within a stream bed by the use of water pressure.

Drift - A horizontal passage underground. A drift follows the vein, as distinguished from a crosscut that intersects it, or a level or gallery, which may do either.

Drift mine – An underground coal mine in which the entry or access is above water level and generally on the slope of a hill, driven horizontally into a coal seam.

Drilling - The process of installing a cylindrical hole in the ground or in the rocks primarily to investigate the subsurface and or collect samples as drill cores

Economic geology - The science of **mineral** deposits, and a study of their characteristics, distribution, mode of occurrence and origin.

Eluvial deposit – Mineral deposit or soil derived from in situ weathering of rocks or gravitational movement or accumulation.

Endowment – The full economic potential of an area due to be mined.

Environmental pollution – The release of toxic substances into the environment.

Epigenetic – Mineral deposit precipitated after the host rock had formed.

Epithermal - Hydrothermal mineral deposits that formed at depths of less than 6km and temperatures of 100 to 200°C.

Explosive - Any rapidly combustive or expanding substance. The energy released during this rapid combustion or expansion can be used to break rock

Extraction - The process of mining and removal of coal or ore from a mine.

Face - The end of a drift, crosscut or stope in which mining work is taking place.

Face drilling - A method used in conventional mining to drill a hole in the coal bed for inserting explosive charges. The holes are drilled into the rock and packed with explosives.

Feasibility study- An assessment of the economic viability of an ore deposit taking into consideration all aspects relevant to mining.

Footwall - The area below where ore is present in a mine.

Gangue – The worthless material surrounds or is mixed with the mineral in ore.

Glory hole - An open pit from which ore is extracted, especially where broken ore is passed to underground workings before being hoisted.

Graben – Elongated depression of the earth's crust lying between two faults.

Grade – The amount of useful minerals or metals contained in a piece of ore.

Greenfield exploration –The search for minerals in an area without previous information or mining.

Hanging wall - The rock or rocks stratigraphically and/or structurally above an ore deposit or the rock on the upper side of an ore deposit.

Haulage - The horizontal transport of ore, coal, supplies, and waste. The vertical transport of ore is called hoisting

Hydrothermal mineral deposit - Deposit that formed from hot aqueous solutions.

Hoist - The machine used for raising and lowering the cage or other conveyance in a shaft Horst – Uplifted and elongated block of the earth's crust lying between two faults.

Host rock - The rock surrounding a mineral deposit.

Hypabyssal – Intrusive igneous rock emplaced at shallow depths within the crust.

Hypogene - Primary minerals in an ore deposit that formed at great depth and high temperature from ascending hydrothermal solutions of igneous origin.

Hypothermal- Hydrothermal ore deposits formed at depths >10km and temperatures between 300 and 500 degrees centigrade.

Incline - Any entry to a mine that is not vertical (shaft) or horizontal (adit).

Industrial mineral - A non-metallic mineral that can be used in construction and manufacturing industries.

Kimberlite- A variety of peridotite; the most common host rock of diamonds

Level - An excavation or passageway driven in the coal, establishing a base from which other workings begin

Lode - A series of associated but closely separated veins that are rich in ore.

Longwall mining - A form of underground coal mining where a long wall of coal is mined in a single slice.

Manhole - A safety hole constructed in the side of a gangway, tunnel, or slope in which miners can be safe from passing locomotives and cars.

Mesothermal - Hydrothermal ore deposits formed at depths of 5 to 10 km and temperatures of 200 - 300 degrees centigrade.

Metallic mineral - A mineral that contains a useful metal.

Metallogeny - The study of the origin (genesis) of ore deposits, particularly in relation to tectonic processes or orogeny.

Metallogenic epoch - A period of geological time favorable for the deposition of mineral deposits characterized by specific deposit types.

Metallogenic province- An area of the earth's crust characterized by a particular assemblage of mineral deposit types.

Metallurgy- The process of separation and purification of metals in ores.

Miarolitic - A term applied to small irregular cavities in igneous rocks, especially granites, into which small crystals of the rock-forming minerals protrude;

Microthermometry – Temperature of small fluid inclusions in ore minerals.

Mineral deposit - An accumulation of useful minerals in a form that has economic potential.

Mineral economics - An academic discipline that focuses on economics and policy issues related to the production and supply of mineral commodities.

Mineral exploration - The systematic search for, discovery and evaluation of economic minerals.

Mineral occurrence- A locality or wherever a mineral of potential economic interest is found, without reference to quantity or quality

Mineral showing - A mineral occurrence that is so small and only of geological interest but not necessarily of economic interest.

Mineral processing - The separation of ore minerals from gangue after mining.

Mineral prospect - Any promising mineral occurrence.

Mineral prospecting - Searching for minerals by visual observation in the field.

Mineral reserves - The portion of indicated mineral resource that is economically feasible to extract.

Mineral resources - The useful substances in the earth's crust that can be of economic benefit.

Mineral resource potential - The probability that minerals will occur in an area.

Mineral zoning – A systematic variation in the composition of a mineral or systematic variation in the distribution of metals or minerals within a deposit, district or region.

Mine - An opening or excavation constructed either above the land surface or underground from which ore minerals are extracted.

Mine development- The operations involved in preparing a mine for ore extraction, which include tunneling, sinking, cross-cutting, drifting and raising.

Mine workings - The entire system of openings in a mine for the purpose of mineral extraction.

Miner - One who is engaged in the business or occupation of extracting ore, coal, precious substances, or other natural materials from the earth's crust.

Mining - The process of extracting valuable minerals from an ore body.

Mining engineer - A person qualified by education, training, and experience in mining engineering.

Mining geology is a branch of economic geology that applies geological knowledge and techniques in the extraction of mineral deposits.

Native elements -Those elements that occur in nature in uncombined form.

Open pit - A mine that is entirely on the surface. Also referred to as open-cut or open-cast mine

Open pit mining- The process of mining an ore body from the surface in progressively deeper steps.

Opencast mining – A method of mining rock or minerals by removing them from an open pit situated close to the surface above ground.

Ore - A mixture of ore minerals and gangue from which at least one of the metals can be extracted at a profit. **Ore body- A**n extensive or continuous mass of ore and gangue distinctly distinguishable by form and character from the enclosing host rocks.

Ore deposit - A mineral deposit that has been tested and found to be of sufficient size, grade, and accessibility to be mined at a profit.

Ore grade- A measure of the concentration of a mineral or metal contained in an ore body.

Ore mineral - A mineral from which one or more metallic elements can be extracted.

Ore reserves - The calculated tonnage and grade of mineralization which can be extracted profitably.

Ore shoot - The portion, or length, of a vein or other structure that carries sufficient valuable minerals to be extracted profitably.

Orogeny - A period of tectonic activity or mountain building characterized by the folding of a portion of the earth's crust.

Oxide minerals - Minerals that contain one or more metals that are combined with oxygen.

Oxidized ore - The alteration of metallic minerals by weathering and action of surface waters that converts them wholly or partially into secondary minerals; oxides, carbonates or sulfates.

Outcrop - An exposure of rock or mineral deposit that can be seen on surface that is, not covered by soil or water.

Overburden - Layers of soil and rock covering an orebody or a coal seam. Overburden is removed prior to surface mining and replaced after the coal is taken from the seam.

Pillar - A column or body of coal left unmined to support the roof.

Placer deposits - Concentrations of ores in gravels usually transported by water.

Porphyry copper - A deposit of disseminated copper minerals in or around a large body of intrusive rock.

Protore - Initial non-economic concentration of metallic minerals that may be economic if enriched by geological processes e.g. weathering.

Quarry - Large deep pit from which stone or other materials are extracted

Raise - A secondary or tertiary inclined opening, vertical or near-vertical opening driven upward form a level to connect with the level above, or to explore the ground for a limited distance above one level.

Ramp - A secondary or tertiary inclined opening, driven to connect levels, usually driven in a downward direction, and used for haulage.

Reclamation – The restoration of land and environmental values to a surface mine site after the ore is extracted.

Reef - Term commonly used in Australia and South Africa to describe a mineralized zone that occurs in a rock layer'

Room-and-pillar mining - A method of mining flat-lying ore deposits in which the mined-out area, or rooms, are separated by pillars of approximately the same size

Run-of-mine - Raw material as it exists in the mine; average grade or quality.

Salting - Act of introducing metals or minerals intentionally or by accident into a deposit or samples, resulting in false assays which may defraud the public or investors.

Shaft - A vertical passageway to an underground mine for moving miners, equipment, supplies and broken ore and waste rock.

Seam - A stratum or bed of coal.

Stratabound deposits - Mineral deposits that lie within sedimentary strata and of sedimentary origin

Stratiform deposits- Mineral deposits laid out like stratified rock layers.

Sedimentary exhalative deposits (SEDEX) - Sediment hosted deposits formed by the discharge of hydrothermal fluids into the sea floor.

Shaft - A vertical excavation connecting surface and mine workings.

Shaft mine – An underground mine in which the main entry or access is by means of a vertical shaft.

Shaft mining - Using shafts driven vertically from the top down into the ground to reach and mine the orebody

Shrinkage stoping - A stoping method which uses part of the broken ore as a working platform and as support for the walls of the stope.

Shortwall – An underground mining method in which small areas are worked (15 to 150 feet) by a continuous miner in conjunction with the use of hydraulic roof supports.

Sludge - Rock cuttings from a diamond drill hole, sometimes used for assaying.

Stope - An excavation in a mine from which ore is, or has been, extracted.

Slope mining - type of underground mining that uses diagonally sloping access shafts to reach the orebodies particularly where drilling vertical shafts may be difficult.

Solution mining – Type of mining where minerals are dissolved into solution and extracted

Stockwork - A large number of closely spaced, planar and irregular veinlets filled with economic and/or gangue minerals.

Stoping - The act of extracting ore from a face or breaking down a stope and excavating it with a pick.

Stripping ratio – the amount of waste rock mined relative to the ore rock mined.

Sump - The bottom of a shaft, or any other place in a mine, that is used as a collecting point for drainage water.

Strip - To remove the overburden or waste rock overlying an orebody in preparation for mining by open pit methods.

Strip mine - An open-pit mine, usually a coal mine, operated by removing overburden, excavating the coal seam, and then returning the overburden

Stripping ratio - The ratio of tons removed as waste relative to the number of tons of ore removed from an open-pit mine.

Sublevel - A level or working horizon in a mine between main working levels.

Sulfide minerals - Minerals that contain one or more metals that are combined with sulfur. Example is chalcopyrite, a copper sulfide.

Supergene - Mineralization and enrichment caused by secondary processes or leaching by descending surface waters.

Surface mine – A mine in which the coal lies near the surface and can be extracted by removing the covering layers of rock and soil.

Syngenetic – Mineral deposits formed at the same time as the host rock.

Tailings - Material rejected from a mill after most of the recoverable valuable minerals have been extracted.

Tenor is the percentage of the ore mineral that is actual metal to be extracted. Often used as synonymous with ore grade.

Tonnage - The quantity of metal or useful mineral in a deposit.

Trench - A long, narrow excavation dug through overburden or blasted out of rock, to expose a vein or ore body during mineral exploration.

Tunnel - A horizontal, or near-horizontal, underground passage, entry, or haulage way, that is open to the surface at both ends. A tunnel (as opposed to an adit) must pass completely through a hill or mount

Underground mine – Also known as a "deep" mine. Usually located several hundred feet below the earth's surface, an underground mine's coal is removed mechanically and transferred by shuttle car or conveyor to the surface.

Underground station - An enlargement of an entry, drift, or level at a shaft at which cages stop to receive and discharge cars, personnel, and material.

Vein - A fissure, fault or crack in a rock filled by minerals that have traveled upwards from some deep source.

Vug - A small cavity in a rock, frequently lined with well-formed crystals. Amethyst commonly forms in these cavities.

Waste - That rock or mineral which must be removed from a mine which has no value.

Wall rock - Rock units on either side of an orebody.

Winze - Secondary or tertiary vertical or near-vertical opening sunk from a point inside a mine for the purpose of connecting with a lower level or of exploring the ground for a limited depth below a level.

Working face is any place in a mine where material is extracted during a mining cycle.

Workings - The entire system of mine excavations.

About the Author

Moses A. Olade is a retired Professor of Economic Geology who taught for more than two decades at several higher institutions including the University of Ibadan, Nigeria, the University of Phoenix, Arizona, Federal University of Technology, Akure and the Achievers University, Owo. He received his B.Sc. (Hons.) degree in Geology from the University of Ibadan, M.Sc. in Economic Geology from the University of Alberta, Edmonton, and a Ph.D. in Applied Geochemistry from the University of British Columbia, Vancouver where he worked on porphyry copper deposits. He served his postdoctoral research fellowship at the Geological Survey of Canada, Ottawa, conducting field work and research on uranium occurrences and base metal deposits. Prof. Olade's professional experience included mineral prospecting and exploration for uranium and base metals in the Northwest Territories and Yukon Canada and California, exploration for gold, lead-zinc and tin deposits in Nigeria, and employment for over a decade in environmental program management in Arizona, USA. He has visited several underground mines in Montana and Idaho, and open-pit mines in Arizona and Nigeria. His research papers on mineral deposits and applied geochemistry are widely published, and he is the author of two recent books: *Mineral Deposits and Exploration Potential of Nigeria*, and *Basics of Economic Geology and Mining Practices*. He was a member of several professional organizations in economic geology and mining, and a distinguished Fellow of the Nigerian Mining and Geosciences Society (FNMGS).

Made in United States
North Haven, CT
09 September 2026

18272893R00102